Practical Sustain

"This book will change how you see the world. [Dra]matic changes are happening to our planet and the only way forward is to [treat] the environment, and each other, with sustainability in mind. By presenting the basic science behind global threats and offering sage advice for doing your part to make a positive difference, this book is a must-have manual for life on Earth in the twenty-first century."

—Professor E. Christian Wells, *Department of Anthropology, University of South Florida, USA*

Robert Brinkmann

Practical Sustainability

A Guide to a More Sustainable Life

palgrave
macmillan

Robert Brinkmann
College of Liberal Arts and Sciences
Northern Illinois University
DeKalb, IL, USA

ISBN 978-3-030-73781-8   ISBN 978-3-030-73782-5 (eBook)
https://doi.org/10.1007/978-3-030-73782-5

© The Editor(s) (if applicable) and The Author(s), under exclusive licence to Springer Nature Switzerland AG 2021

This work is subject to copyright. All rights are solely and exclusively licensed by the Publisher, whether the whole or part of the material is concerned, specifically the rights of translation, reprinting, reuse of illustrations, recitation, broadcasting, reproduction on microfilms or in any other physical way, and transmission or information storage and retrieval, electronic adaptation, computer software, or by similar or dissimilar methodology now known or hereafter developed.

The use of general descriptive names, registered names, trademarks, service marks, etc. in this publication does not imply, even in the absence of a specific statement, that such names are exempt from the relevant protective laws and regulations and therefore free for general use.

The publisher, the authors and the editors are safe to assume that the advice and information in this book are believed to be true and accurate at the date of publication. Neither the publisher nor the authors or the editors give a warranty, expressed or implied, with respect to the material contained herein or for any errors or omissions that may have been made. The publisher remains neutral with regard to jurisdictional claims in published maps and institutional affiliations.

Cover credit: Icons by Cokolfly

This Palgrave Macmillan imprint is published by the registered company Springer Nature Switzerland AG.
The registered company address is: Gewerbestrasse 11, 6330 Cham, Switzerland

Acknowledgments and Dedication

When putting together a book like this there are many people to thank. First of all, big thanks to the team at Palgrave Macmillan for their support on this and other projects. I am always grateful for the level of excellent support and thoughtful advice. Also, sincere thanks to the reviewers who provided helpful feedback that made this work richer and more relevant. None of this work would have been possible without the kindness of my colleagues at the University of Northern Illinois and my friends across the world—particularly those in Long Island, New York, and Florida. Finally, big thanks to my family who have always been there for me—especially Mario and Elis Gomez, Patricia and Nelson Sohns, John Brinkmann, and Sharon and Robert Hoff.

This book is dedicated to everyone who tries each day to live a greener life. We cannot be perfect in our pathway toward sustainable living. We all have an impact on our beautiful Earth. All we can do is to try to do the best that we can.

Contents

Part I	Defining Our Great Global Challenges	1
1	Change Yourself and Change the World	3
2	Our Climate Change Challenge	13
3	Our Great Sustainability Challenge	37
4	Our Ethical Responsibility	67
Part II	Tackling Climate Change	85
5	How You and Your Family Can Reduce Your Carbon Footprint	87
6	How Your Community Can Reduce Its Greenhouse Gas Impact	103
7	How Your School, Non-profit Organization, or Business Can Reduce or Eliminate Its Carbon Footprint	121

Part III	Environmental Sustainability	139
8	Moving to Green Energy	141
9	Protecting Our Water Resources	159
10	Dealing with the Garbage Around Us	177
11	Saving Ecosystems	191

Part IV	Building Just and Equitable Economic and Social Systems	207
12	Building a Just and Sustainable Society	209
13	Green Your Economy	227
14	Sustainable Travel and Leisure	243
15	Tune Out, Buy Nothing, and Get Educated	261

References	277
Index	291

List of Figures

Fig. 1.1	Places like New York have well-known sustainability initiatives. What types of sustainability agenda is in place in your community?	5
Fig. 1.2	Our beaches are optimistic places that seem to be natural, but we often rely on engineering to keep them in place through jetties, beach renourishment programs, and breakwaters. As a result, they are metamodern places that are not in tune with the environment	9
Fig. 2.1	Greenhouses use the greenhouse effect to help trap heat. This greenhouse near Amsterdam is warm enough to keep tropical orchids comfortable in a northern European climate where they would not survive	15
Fig. 2.2	Philadelphia, Pennsylvania, is a member of ICLEI which helps the city benchmark its sustainability initiatives with other cities around the world. Is your community a member?	28
Fig. 3.1	Nature is highly altered in the Anthropocene as a result of human activity. Signs of alteration can be found everywhere including in this image in France	39
Fig. 3.2	Our quest for more and more luxury goods has a direct impact on the environment	49

List of Figures

Fig. 4.1	This old railroad track has been turned into a walking trail and some aspects of a natural ecosystem have returned. How we use land over time has an impact on the environment	70
Fig. 4.2	This prairie preserve in Illinois has been intentionally restored to bring back the native prairie ecosystem. The restoration shows that people care about the environment and our interaction with it	77
Fig. 5.1	Calculating your greenhouse gas footprint can be complicated. For example, the people living in these condominiums and apartments along this Vancouver waterfront have to consider things like heating, cooling, lighting, transportation, waste, and food sources	90
Fig. 6.1	Egypt has notoriously problematic air pollution problem caused by a variety of emissions including car exhaust and local cooking fires	107
Fig. 6.2	Every community can calculate their greenhouse gas emissions including those associated with festivals like this Oyster Festival in Oyster Bay, New York	112
Fig. 7.1	Oxford University has tried to embrace sustainability in big ways. How have schools in your region embraced sustainability?	127
Fig. 7.2	Ford Motors physically shows their commitment to sustainability in their green roof and with extensive arrays of solar panels. Ford is also transforming many of their car models to electric vehicles	132
Fig. 9.1	Niagara Falls shows the water cycle in all of its glory	160
Fig. 9.2	The dry uplands near San Antonio go through boom and bust water cycles. As a result, the Edwards Aquifer Authority helps to manage water within the Edwards Aquifer watershed	174
Fig. 11.1	Florida's wetland ecosystems contain a tremendous amount of biodiversity, even though they have been heavily altered by human activity	197
Fig. 11.2	Urban ecosystems, such as this one on my campus in DeKalb, Illinois, are home to a fascinating array of plants and animals—and humans	199
Fig. 12.1	The world has struggled with maintaining human rights around the world. What can you do in your community to try to make sure that human rights are protected?	211

Fig. 12.2	Not all people have access to fresh and healthy food such as is found in this market in Alexandria, Egypt	218
Fig. 13.1	This small organic farm is an example of a green business	229
Fig. 13.2	New York City is home to the United Nations and many Fortune 500 companies like BlackRock that are making the world's economy greener	235
Fig. 14.1	Cruising has gained in popularity in recent years, but some cruising is greener than others. This small cruise ship on the Nile, called a dahabieh, uses wind for part of its power	249
Fig. 14.2	Adventure tourism is an exciting form of travel that can take you in new directions. But how green is it?	254
Fig. 15.1	A group of community volunteers installing an osprey nest structure on the shores of Long Island Sound	262
Fig. 15.2	Celebrating Earth Day, like these students, helps to build community and participants can share new sustainability strategies	272

List of Tables

Table 5.1	Simple, hard/expensive, and innovative/life-changing approaches to energy reduction in the home	95
Table 5.2	Simple, hard/expensive, and innovative/life-changing approaches to reducing greenhouse gases from our transportation choices	97
Table 5.3	Simple, hard/expensive, and innovative/life-changing approaches to reducing greenhouse gases from our food choices and personal waste management	99
Table 6.1	Goals of the Louisville, Kentucky, sustainability plan (LouisvilleKY.gov, 2020)	105
Table 6.2	Definitions of Scope 1, Scope 2, and Scope 3 emissions	111
Table 6.3	Major steps in a community-based greenhouse gas inventory	113
Table 6.4	Simple, hard/expensive, and innovative/life-changing approaches to reducing greenhouse gas emissions in a community from energy production, transportation, and buildings	118
Table 7.1	Simple, hard/expensive, and innovative/life-changing approaches to sustainability in schools	124
Table 7.2	Simple, hard/expensive, and innovative/life-changing approaches to reducing greenhouse gas emissions in three different types of non-profits: youth groups, adult service groups, and community service organizations	128

Table 7.3	The NGO Climate Compact (taken from https://www.interaction.org/wp-content/uploads/2020/04/Climate-Compact.pdf)	131
Table 7.4	Simple, hard/expensive, and innovative/life-changing approaches to reducing greenhouse gas emissions in business	133
Table 8.1	Top coal-consuming countries in 2019 (Enerdata, 2020a)	142
Table 8.2	List of top crude-oil producers and petroleum product consumers in the world as of 2019	144
Table 8.3	List of top natural gas producers and natural gas consumers in the world as of 2019 (Enerdata, 2020b)	146
Table 8.4	Production of renewable energy in 2019. Note that hydroelectric energy is the largest source of renewable energy and that wind energy produces twice as much electricity as solar energy (Our World in Data, 2020b)	150
Table 8.5	Top solar energy-producing countries (Our World in Data, 2020b)	153
Table 8.6	Simple, hard/expensive, and innovative/life-changing approaches to infusing green energy in your life	155
Table 9.1	The list of the countries with the highest annual per capita water withdrawal (Our World in Data, 2020)	162
Table 9.2	Simple, hard/expensive, and innovative/life-changing approaches to conserve water and reduce water pollution	173
Table 10.1	List of common household hazardous wastes	179
Table 10.2	List of major classes of recycled materials with examples	184
Table 10.3	Simple, hard/expensive, and innovative/life-changing approaches to reduce waste and live a minimalist lifestyle	189
Table 11.1	Major types of wetlands	194
Table 11.2	Simple, hard/expensive, and innovative/life-changing approaches to protecting and preserving ecosystems	205
Table 12.1	Articles of the Declaration of Human Rights	213
Table 12.2	Simple, hard/expensive, and innovative/life-changing approaches to advancing environmental justice	224
Table 13.1	Simple, hard/expensive, and innovative/life-changing approaches to advancing a green economy	240
Table 14.1	Simple, hard/expensive, and innovative/life-changing approaches to greening your travel and vacations	258

Part I

Defining Our Great Global Challenges

This book is designed to help readers better understand sustainability and the actions that they can take to make the world a more sustainable place that will provide clean and healthy environments for generations to come. Right now, we are not on this path and instead cause serious environmental problems that range from climate change to environmental racism. The chapters in this part of the book review background on sustainability. After an initial chapter that defines sustainability and summarizes the outline of the book, Chap. 2 takes a deep look at the science of climate change and why we should be actively engaged in trying to mitigate its impacts. Of course, climate change is not the only sustainability problem and Chap. 3 highlights the environmental, social, and economic issues associated with sustainability by reviewing some of the most vexing issues the world is facing. The final chapter in this section, Chap. 4, looks deeply at environmental ethics and how we need to infuse environmental ethics within our culture in order to help support the changes we need to make to ensure that our planet can be sustainable for generations to come.

1

Change Yourself and Change the World

Introduction

Our beautiful world is facing huge existential challenges as a result of our human activity and it is time we did something about it. The challenges are great. We are experiencing unprecedented global climate change, we have widespread pollution choking our waters, many areas are recording rapid species and ecosystem decline, and environmental racism is on the rise—and this is just the tip of the melting iceberg. Scientists have been recording these changes for the last several decades and have been ringing the warning bell louder and louder and it often seems as if no one is listening. We continue, as individuals, organizations, and governments, to act as if the world had limitless resources and as if the world could absorb the many abuses we inflict on the planet. We are finding out very quickly that there are indeed limits to what the earth can take.

This book focuses on how you and your organizations—non-profit organizations, businesses, community groups, and governments—can make fundamental changes in order to live more sustainably and in harmony with nature. After some background on our great sustainability challenges such as climate change, the book dives into specific themes of sustainability in order to provide concrete suggestions as to how to make specific changes with a goal of living a more carbon neutral life and one that makes a positive contribution to the three E's of sustainability: environment, equity, and economy.

This book was inspired by a project of mine called the Thirty-Day Sustainability Challenge which focuses on challenging groups of people over 30 days to make fundamental lifestyle changes. Each day of the 30 days has a different theme such as organic food, transportation, or environmental racism. Each time I run the challenge, dozens of people from around the world participate in trying to transform their lives through this sustainability coaching exercise. I run the groups on Facebook or Teams and go live each day to provide short five-minute mini-coaching sessions on the day's topic. Participants are able to comment live while I am speaking and I can react in real time to their thoughts and questions.

In running the Thirty-Day Sustainability Challenge, I realized that there was a hunger for greater information about sustainability and what individuals or groups of people could do to make a greater difference. This book is the outcome of that realization. I believe that most people genuinely want to do the right thing, but do not know how to begin a transformation into a greener life. This book, like the Thirty-Day Sustainability Challenge, seeks to provide guidance and suggestions to help people on their path toward sustainable living.

At the outset, it is important to recognize that no one is living a 100% sustainable life. Many look at the sustainability or green community as judgmental toward others or find people like me who advance a sustainability agenda but live in imperfect sustainable lifestyle as hypocritical. We might feel judged about the kind of car we drive, the food we eat, the house in which we live, and the clothes we wear. I find this unfortunate. Everyone can live a greener life and we should not judge others for where they are on their path. Someone with a giant SUV may have a reason for having the vehicle we may not understand or they may be paying it off so that they can trade it in on an electric car. The bottom line is that I look at sustainability as a no-judgment zone. You are starting where you are starting and no matter your ecological footprint, I welcome you to the sustainability community. We need everyone on board to make a change.

I think it is important to look at sustainability within a big tent. In the United States, due to the lack of national leadership on the issue, cities have taken much of the leadership role on sustainability. Thus, places like Chicago, New York, and Louisville are places where sustainability is part of the fabric of governance (Fig. 1.1). But, if you go outside of the borders of these communities, sustainability is not a major issue and is rarely ingrained in government or community conversations. This is a significant loss because cities are only a piece of the United States and many regions are being left behind and there are many people who could do much more to help. What is important to recognize is that most people want to help. We are in a time when we are past any serious debate on the realities of climate change and broad human-induced environmental degradation. We can all see it and we all want to do more.

Sustainability is defined broadly as development that meets the needs of the present without limiting the ability of future generations to meet their needs (Brundtland, 1987). However, the idea of sustainability has split into two broad areas in recent years (Brinkmann, 2020). One area focuses on things like the development of lesser developed nations and the other area focuses on how developed nations can lower their overall

Fig. 1.1 Places like New York have well-known sustainability initiatives. What types of sustainability agenda is in place in your community?

ecological footprint. In the former example, developing nations are seeking to do things like improve water quality and access, provide electricity to communities, and improve public health. In the latter example, developed nations are seeking to advance renewable energy sources, create organic farms, and address environmental equity issues. The challenge with all of these initiatives is that some are more effective than others at addressing sustainability and some may even put us backward.

In my opinion, because of these challenges, sustainability is one of the first metamodern disciplines. Metamodernism is both a philosophy and a social movement that emerged in the last decade or two from postmodernism which itself was a reaction to modernism (van den Akker & others, 2017). Modernism can be looked at as more or less the outgrowth of the enlightenment. Our framework of science and even education itself is a product of modernism. In the 1980s and 1990s, postmodernism led to the questioning of modernism and sought to deconstruct the frameworks of modernism to understand, in part, power structures in modernism and to reemphasize the unemphasized.

To digest down postmodernism to an essence, there is no truth. Instead, there are constructions of reality built by those in power to create an understanding of our world. The movie *The Matrix* nicely creates the feeling of postmodernism by totally deconstructing reality to create a new reality that may or may not itself be real. The Warhol prints of Mao and Marilyn Monroe also deconstruct the meaning of art and image and the singer Madonna constantly reinvented her image since the 1980s in an embodiment of postmodernity—there is no Madonna, just an annual recreation of identity.

Metamodernism is a reaction to postmodernism in that it recognizes the realities of both modernism and postmodernism. It seeks to take an optimistic approach to the world by recognizing that both modernism and postmodernism exist at the same time. The video, "This Is America," by Childish Gambino, and the *Borat* movies are cultural representations of metamodernism that provide both real and unreal experiences. It isn't magical thinking. The reality of the unreality actually exists. Politically this was represented by the Trump administration and the very metamodern policies associated with COVID. The US government was both working on the pandemic while not working on it. It will go away, it is a hoax, it is killing thousands, and it is a real public health crisis.

So why is sustainability a metamodern discipline?

I have written a great deal about the different ways that sustainability is enacted, discussed, and implemented around the world. I sometimes talk about it as "a tale of two sustainabilities" whereby one sustainability is the sustainability of the West and the other is the sustainability of the developing world (Vercoe & Brinkmann, 2012). In other places I have written about surfing and suffering sustainability where Western approaches are cool, optional, and quantitatively of comparatively limited value (due to huge per capita consumption) and where suffering sustainability is about existentialism and improving challenging situations. Now I think a better way of framing sustainability is within metamodernism.

Here are two examples of what I mean from the recycling and energy worlds.

If we look at plastic recycling in the United States, we know that the vast majority of the plastic is shipped overseas and of that plastic the vast majority of it is diverted to local waste streams or litter and is not recycled. We have built a whole infrastructure around plastic recycling but it is of limited actual value. Of course, the best thing we could do is not use plastics. But instead, we institutionalize recycling within a sustainability modernist approach even though there are metamodern realities that show that it is of limited value. Should we stop plastic recycling? No. But is plastic recycling a problem? Yes. Do we feel optimistic about recycling overall? Yes. Metamodern.

Green energy is another metamodern issue. We have rapidly ramped up the development of green energy across the world. That is truly a wonderful thing. However, during the same time that we have ramped up these energy sources, global consumption of oil and natural gas, two important drivers of climate change, have increased substantially and the overall world's energy use has gone up and continues to rise. Thus, green energy has not replaced oil and natural gas consumption at the global level. It has only added to the overall amount of energy that is being produced around the world. Do we want to continue to produce green energy? Yes. Has green energy led to a reduction of the global use of fossil fuels? No. Do we feel good about wind and solar? Yes. Metamodern.

The field of sustainability is inherently a quantitative one. We measure a baseline, develop policies to enact change, and then measure the results of the outcome. To some, however, sustainability is about the "feels" of the environment. It feels right to recycle and it feels right to drive a Tesla. However, the feels often mask broader issues with overall consumption. We don't always question whether we needed to purchase the plastic or buy the car in the first place.

When you look closely at the discourse around sustainability, you can find metamodernism everywhere. As a result, I think it is important for sustainability initiatives in the coming years to focus on authenticity and quantitative approaches that clearly demonstrate impacts locally, nationally, and globally. We also have to recognize connections. Solving a plastic waste problem in a community in the United States could create a burden for communities in Africa. Adding green energy without serious reductions in carbon-based fuels globally does not really impact our global climate change problem.

I doubt that the metamodern issues we are facing in our society will go away soon. We want to be optimistic about a variety of sustainability issues (Fig. 1.2). We think we can manage our way through climate change and most of us live our lives without changing our behavior. It is the optimism that leads to metamodernism. This book focuses on the optimistic real changes that we can all make to help to make the world a better place. It seeks to challenge readers into making changes that will lead to authentic measurable impacts on the planet.

This book is divided into four parts. Part I, which includes Chaps. 1, 2, 3, and 4, highlights background on sustainability and makes a case for why we should care about the issue. After this introductory chapter, Chap. 2 focuses on climate change, one of the most challenging issues we are facing at this moment. It highlights much of the scientific evidence for climate change and then reviews some of the significant work being done by the Intergovernmental Panel on Climate Change which is organized by the United Nations. Chapter 3 delves deeper into other sustainability challenges we are facing within the three E's of environment, equity, and economy. Since the term sustainability was first emphasized

1 Change Yourself and Change the World

Fig. 1.2 Our beaches are optimistic places that seem to be natural, but we often rely on engineering to keep them in place through jetties, beach renourishment programs, and breakwaters. As a result, they are metamodern places that are not in tune with the environment

by the United Nations in 1987, we have developed a keener sense of the major issues we are facing around the world and how they are interrelated. No longer can we examine one issue from the lens of just the environment. We need to understand how social and economic systems are impacted by problematic environmental issues across the globe. Part I concludes with a discussion of broad environmental ethics in Chap. 4. This chapter delves into why we should care about sustainability and how care for our planet is an evolution of human ethical thinking.

Part II focuses specifically on what we can do to address the growing problem of climate change. Each remaining chapter of the book concludes with a summary of simple, hard/expensive, and innovative/life-changing transformations you can make to live a more sustainable life. Chapter 5 examines how you as an individual can reduce and even eliminate your carbon footprint. There are many options for any of us to make serious changes to live a greener life and we need to take advantage of them. At the same time, all kinds of climate change initiatives are available for local, state, and national governments. Chapter 6 reviews these initiatives and provides suggestions on how you can help make a difference where you live. The final chapter in Part II, Chap. 7, provides guidance on how businesses, non-profits, and schools can reduce their greenhouse gas footprint. The goal of Part II is to help move our decision-making in all aspects of our lives toward significantly reducing our greenhouse gases.

Part III moves away from the important issue of climate change to look at how we can make a difference in other aspects of environmental sustainability within four distinct themes in four separate chapters. Chapter 8 provides background on the kinds of energy we are using and how we can use less of it and embrace a green energy economy. We are quickly moving more and more toward renewable energy, but are still reliant on dirty forms of energy. Pollution problems, which include a range of emerging pollutants like plastic and pharmaceuticals, are the foci of Chap. 9. We release a variety of pollutants, like nutrients, pesticides, and heavy metals, into the environment where they can contribute to serious public health problems. This chapter highlights how we can stop contributing to the global pollution crisis. Chapter 10 engages on waste and garbage, which are very closely tied to our consumptive habits. Finally, Chap. 11, the last chapter in Part III, takes on the important topic of ecosystems and how we can do more to protect them. All over the world ecosystems like coral reefs and tropical rainforests are under threat and we need to find ways to protect them.

Part IV takes on the economic and equity E's of sustainability by focusing on building just and equitable social systems—key themes that are central to our modern definition of sustainability. Chapter 12 begins this final part of the book by focusing on building a just society. The chapter looks deeply at issues of human rights, environmental racism, and environmental justice and provides guidance as to how we can all engage on these important issues. In contrast, Chap. 13 brings forward the idea of how economic systems embrace sustainability and how we can all be part of economic transformations that will help green local, regional, national, and international economic systems. Chapter 14 focuses on one important economic system that is a big part of our lives, travel and tourism. Most of us take part in some form of travel and tourism activities throughout the year and these activities can have a big ecological footprint. The chapter reviews the impacts of tourism, how we can make greener choices in travel, and how we can take different types of vacations that have less of an impact on the planet. The final chapter of Part IV, and of the book, confronts the issues of consumerism, leadership, and education. It provides a call for action to make sustainability a larger part of our day-to-day decision-making and challenges us all to be sustainability leaders that support a range of educational initiatives that will help make our planet habitable for generations to come.

In approaching this book, the reader should know that they will be challenged to make different lifestyle choices. There are easy choices that can be made that anyone can do. These are the low-hanging fruit of sustainability that make a big difference and that get us partway there. Then, at the next level, there are hard or expensive choices that you can make that get you a bit closer to living more sustainably. Finally, there are innovative and life-changing transformations that you can make that make fundamental changes that help you live in greater harmony with nature. Perhaps the first time you read this book you will be comfortable in making some of the easy changes to kick-start your path toward sustainability. With a second or third read, you may want to make deeper and more meaningful changes to how you live your life. No matter how far you take yourself down the path to sustainable living, the point is to get engaged with the issue to help start the global transformation needed to help us get off of the unsustainable path we are on.

We all know that unless we make significant, meaningful changes that we will continue to have changing climates and deteriorating environments as a result of human activity. We also know that we have the power to stop climate change and solve key environmental, social, and economic problems that are damaging our planet. We all are responsible for the problems and we all need to be part of the solution. I hope that this book helps guide us on a better path to sustainable living. By transforming even one small piece of your life you are making a difference. It is my hope that one of your lifestyle changes will lead to another so that you can live in greater harmony with the planet. As many have noted, we only have one planet. We need to do all that we can to make sure that it is around for generations after us. It is the right thing to do.

References

Brinkmann, R. (2020). *Environmental sustainability in a time of change*. Palgrave Macmillan.

Brundtland, G. H. (1987). *Report of the World Commission on Environment and Development: Our common future*. https://sustainabledevelopment.un.org/content/documents/5987our-common-future.pdf

Van den Akker, R., Gibbons, A., & Vermeulen, T. (Eds.). (2017). *Metamodernism: Historicity, affect, and depth after postmodernism*. Rowman & Littlefield Publishers.

Vercoe, R., & Brinkmann, R. (2012). A tale of two sustainabilities: Sustainability in the global north and south to uncover meaning for educators. *The Journal of Sustainability Education*. http://www.susted.com/wordpress/content/a-tale-of-two-sustainabilities-comparing-sustainability-in-the-global-north-and-south-to-uncover-meaning-for-educators_2012_03/

2

Our Climate Change Challenge

Introduction

In the summer of 2019, a group of Pacific nations met in the island state of Tuvalu to discuss climate change (Packham, 2019). Many of the small island nations in the Pacific basin face an existential crisis given that the world has yet to come together to try to solve the world's climate change problems. For years, places like Tuvalu have tried to get the world's attention about the severity of the problem. Whole nations can disappear (Connell, 2016). Tuvalu, which has a maximum elevation of only a few meters, is one of the most at risk places. Some areas of Tuvalu and other Pacific Island countries are already dealing with rising sea levels in their most low-lying areas (Constable, 2016). Others are feeling the impacts of the shifts of climate and the concomitant climate extremes of too much rainfall or too little rainfall and enhanced tropical storms.

As the nations were meeting, they discussed the latest scientific reports about the evidence of climate change and the recommended changes to national policies. However, one country, Australia, rejected some of the main efforts to try to reduce greenhouse gases, specifically the development of new coal mines and the efforts to reduce coal consumption by moving to green energy.

Why did Australia try to water down the agreements at Tuvalu? It has an abundance of coal that it exports and it relies on coal as part of its diversified economy. Australia, like many countries of the world, is taking a short-term pragmatic stand for its own benefits. It is rejecting the concerns of countries that are most at risk to the impacts of climate change. In many ways, Australia is acting like many organizations that do not change their behavior to reduce the greenhouse gases that cause climate change. In fact, Australia is just like many of us in that we as individuals are not making decisions about our own greenhouse gas production. We have the same moral responsibility as Australia in taking action. When we take long international flights or drive a gas-guzzling car, are we any different from Australia? Or when our companies or governments make bad environmental choices, aren't they too like Australia? We can all do better. The evidence for climate change is all around us but many of us look the other way.

This chapter will review the science of climate change and outline the types of evidence we have for it happening right now all around us. It will also summarize some of the likely impacts we will see in the coming decades and centuries if we choose to do nothing about the problem. It will also review the kinds of efforts, like the Paris Climate Accord, that we have made so for. As we will see, we have a long way to go if we are going to try to reduce the impacts of climate change on our planet.

The Science of Climate Change

Most of us have walked into a greenhouse on a cool day (Fig. 2.1). Growing up in Wisconsin, I loved going to our local nursery in the early spring to pick out some annuals that we would plant in some of our border beds. I would ride with my older sister in her car and when we arrived at the garden center, the cool air would hit us as we walked out of her 1969 Mustang. We quickly removed our coats when we entered the humid greenhouse. Here, the air was warm and moist and small beads of sweat would form on our foreheads as we picked out the pansies we

Fig. 2.1 Greenhouses use the greenhouse effect to help trap heat. This greenhouse near Amsterdam is warm enough to keep tropical orchids comfortable in a northern European climate where they would not survive

wanted. The greenhouse did not have a heater. Instead, it gained its heat from the sun.

The reason that it was warmer in the greenhouse than outside of it is because of the phenomena aptly named the *greenhouse effect*. You don't have to experience a greenhouse to feel its impacts. Most of us have had the experience of entering a car on a sunny day. During cold snaps, it is a pleasure to enter a car that is warmer than the outside temperature. However, during warm months, the temperatures can be extreme and even life threatening if helpless children or animals are left in cars without air conditioning in the summer.

The way that the greenhouse effect works is that incoming solar radiation contains energy that heats the atmosphere in the greenhouse or car (Anderson et al., 2016). That heat is absorbed by molecules like water and carbon dioxide and thus becomes trapped within the greenhouse or car, thereby causing the temperature to rise. Moister environments with lots of water molecules, like what we find in a greenhouse, can absorb lots of heat and become quite warm.

The earth's atmosphere is like a giant greenhouse. As solar radiation comes into it, molecules like water, carbon dioxide, and methane, can absorb tremendous amounts of heat. In fact, if we did not have these molecules in our atmosphere, the earth would have a very different temperature regime. The key, of course, is to have a good balance of heat-absorbing chemicals so that the atmosphere does not get too hot or too cold. If we have too few of these molecules the earth can cool down. If we have too many, the earth can heat up. This cooling and heating can alter global climate systems to the point that it can change entire ecosystems (Hoegh-Guldberg et al., 2007).

The challenge, of course, is that we have increased the amount of greenhouse gases into the atmosphere, particularly carbon dioxide and methane (IPCC, 2014). Carbon dioxide is emitted when we burn fossil fuels like coal and petroleum (EPA, 2019). These materials formed in the earth over millions of years during the Mesozoic Era from decaying plants and animals. Over the last 150 years, we have released the carbon that took millions of years to form. It is like a giant gas bubble of carbon dioxide suddenly popping in our atmosphere. Methane, otherwise known as natural gas, is often associated with fossil fuels. It is also released as organic matter decomposes in swamps and landfills. Methane also escapes from small leaks in pipelines that bring vast quantities of methane to power plants, factories, and homes (Heede & Oreskes, 2016).

The problem with methane is that it is a far more potent greenhouse gas than carbon dioxide. Because carbon dioxide is such a common greenhouse gas, all other greenhouse gases are compared to it using units called carbon dioxide equivalents. Carbon dioxide obviously has a carbon dioxide equivalent of 1. However, methane has a carbon dioxide equivalent of 25. This means that methane is 25 times more powerful of a greenhouse gas than carbon dioxide. There are other less common greenhouse gases such as chlorofluorocarbons that have carbon dioxide equivalents hundreds of times that of carbon dioxide (IPCC, 2018).

The United States National Oceanic and Atmospheric Administration has been monitoring gases in the atmosphere at a number of locations for decades (Global Monitoring Laboratory, 2020). One of the most important of these observatories is high on the mountain of Mauna Loa in Hawaii. Over the years, they have found that carbon dioxide and methane have been steadily increasing. As I am writing this today, the carbon dioxide levels are over 415 ppm (parts per million). Last year at about the same time the concentration was about 411 ppm. Ten years ago, it was 390 ppm. Back in the 1960s it was around 315 ppm. Methane has also trended up. Today, methane levels in the atmosphere are around 1877 ppb (parts per billion). Last year, the concentration was 1866 ppb. Back in the 1980s the atmosphere had about 1630 ppb of methane.

It is important to stress that these are unprecedented levels. At no time in human history have we had such high levels of greenhouse gases in the atmosphere. Michael Mann and his colleagues several years ago noted that the rise in atmospheric greenhouse gases is clearly tied to increases in temperature over time (McIntosh, 2019). These authors published the famous hockey stick graph that helped to bring great awareness to the issues we were facing as a society. While this work was important, it was not the first important work to demonstrate that we were on a climate collision course. Many authors going back generations warned us about the impacts of greenhouse gases into the atmosphere. They noted that the science was clear: the more greenhouse gases that are emitted into the atmosphere, the warmer the atmosphere will become.

It is crucial to highlight that climate is different from weather. We all experience weather fluctuations that occur day to day or year to year. These are short-term changes that are normal and common. Long-term patterns, such as the average annual temperature or rainfall using data going back decades, are how we understand climate. Climate scientists study the long-term patterns such as temperature, precipitation, atmospheric wind patterns, and severe storms in order to understand our earth's systems. These patterns have evolved during the last several thousand years when we have had a relatively uniform atmospheric chemistry—including the presence of greenhouse gases. Now we are starting to see a number of changes to daily weather patterns which is indicating that we are now in a time of change. It is important to note that climate is always changing with or without us. What is happening now is that we are driving it in a particular direction toward warming and this is starting to cause some chaotic weather conditions.

Climate Change Denial

In the past few decades some scientists have questioned whether or not human-induced climate change is real. However, as the evidence has poured in, there are only a handful of scientists who question the broader scientific consensus. Indeed, it is hard to find a major climate scientist who denies the basic framework of climate change that was outlined in the last section. So why is so much attention given to climate change denial?

The main reason is that there are very powerful economic interests that will lose if we move away from traditional fossil fuels (Tucker, 2012). A variety of interests have funded advertising campaigns and supported very vocal politicians who espouse climate change denial viewpoints. Currently, there are important legal actions being taken in New York and Massachusetts. In the latter case, Massachusetts is suing Exxon over what they claim are deceptive practices around climate change (Larson, 2020). The lawsuit alleges that Exxon knew about climate change, but engaged in a variety of deceptive practices to dupe investors around the risks the company faced around climate change. The lawsuit in New York is following similar trends as the famed tobacco lawsuits of the 1990s in that the state is claiming that energy companies knew about the role that their products played in causing climate change but worked to deceive the public and their shareholders around climate science.

This deception was supported by politicians who were supported by the fossil fuel industry (Franta, 2018). Many politicians from energy-rich countries or regions, such as Scott Morrison from Australia, Donald Trump from the United States, and Jair Bolsonaro from Brazil are all strong advocates for the fossil fuel industry and have worked against climate change policy in their countries and in international fora. As the public has become more sophisticated about climate change in the last several years, these anti-climate positions are becoming more difficult to maintain. The public is experiencing climate disruption in real time. The 2019 fires in Australia (Boer et al., 2020) and the slow and steady inundation of low-lying areas in places like Miami (Jacobson & others) and the low-lying islands of the Pacific Ocean provide clear evidence to anyone paying attention to the news that we are in a time of change.

Climate Change Evidence

In the 1980s, as scientists began to be concerned about the impacts of greenhouse gases on our planet's climate, the World Meteorological Organization established a working group called The Intergovernmental Panel on Climate Change (IPCC) to assess the published scientific information on climate change and its risk to our planet. The IPCC consists of some of the top scientific minds in the world who scour the scientific literature to understand what we know about this important topic. The IPCC team is broken into three working groups. Working Group I looks at the science of climate change. Working Group II focuses on the vulnerability of our socioeconomic and natural systems to climate change. Finally, Working Group III assesses how to solve the climate change problem by looking at how to reduce emissions and mitigate climate change.

Since the organization formed, it has published five assessment reports that summarize their findings. They come out every five to seven years. The last assessment report came out in 2014 (IPCC, 2014) and the next one is due out in 2022. The reports consist of three main areas that follow the charges of the three working groups. While it is impossible to review all of the findings of the various reports in this space, it is worth highlighting some of the key conclusions around the evidence of climate change in the most recent report. The science is clear. The climate is changing across the planet as a result of greenhouse gas pollution.

- Climate change is unequivocal.
- Observed changes are unprecedented over decades to millennia.
- The atmosphere and oceans have warmed.

- Snow and ice have diminished.
- Sea level has risen.
- Atmospheric greenhouse gas emissions are the highest in our history and have not been seen at this level for at least 800,000 years.
- Extreme weather and climate events (decrease in cold temperature extremes, an increase in warm temperature extremes, increase in high sea levels, and increase in the number of heavy precipitation events) are occurring at more regular intervals.

The report provides an abundance of data to support these findings including maps and graphs from the scientific literature. The report also predicts what we can expect to experience in the coming decades if we do not reduce our greenhouse gas emissions. The results are stark and we have a difficult future ahead of us.

- We can expect to see further warming and long-lasting changes in climate.
- We will see the likelihood of severe, pervasive, and irreversible impacts on people and ecosystems.
- The key driver of global climate change will continue to be greenhouse gas emissions.
- It is very likely that heat waves will occur more often and last longer.
- Extreme precipitation events will become more intense and frequent.
- The ocean will continue to warm and acidify.
- Global mean sea level will continue to rise.
- Climate change will amplify risks and create new risks for livelihoods and for food and human security.
- Even if we stop increasing the use of greenhouse gases, the impacts of global climate change will continue for centuries.

These key findings are based on a variety of models and scenarios that show that global climate change will significantly disrupt our way of life in the coming years. What is particularly concerning is that the impacts will continue on for decades or centuries even if we stop greenhouse gas pollution. We will have to make significant changes now in order to ward off long-term problems.

The IPCC predicts that the impacts of climate change will not be evenly distributed. Low-lying coastal areas will obviously be most impacted due to rising sea levels. In addition, these places will be impacted by the likely increase in coastal and tropical storms. Growing zones will change thereby causing disruptions in agricultural systems. I was recently in Minnesota, one of the most important agricultural regions in the United States, where climate scientists are predicting that this state will soon have the climate of Missouri. While this may seem like that is a small change, in reality, it would be a profound change for the farming community who are used to growing crops under particular temperature and rainfall regimes. This, combined with likely ecosystems disruptions across the continents, will ensure that all areas of the planet will experience climate change in some way. Given the recent experience with COVID-19, many public health experts are very concerned about the movement of diseases, especially tropical diseases, as regions warm.

Of course, not all people are able to adapt to change in the same way. Research has shown that the most vulnerable people in any type of natural disaster, such as hurricanes or floods, are the poor—particularly those without strong social networks (Cutter, 2006). Wealthier individuals often find ways to adapt to change. It is important to consider what types of social disruption may occur as countries all over the world try to adapt to climate change in the coming decades. There are large cities, some rather impoverished, in some of the lowest-lying areas of the world. How will these places adapt as they become stressed or inundated?

As we start to experience the accelerating impacts of climate change, how will we react? Will we adapt to a new way of life, or will conditions lead to conflict and loss of life? In some ways, we are already adapting and becoming more resilient. Some communities have developed plans for dealing with a new normal. They have limited low-lying and coastal area real estate development and have taken steps to protect infrastructure. Some places are working on creating more resilient social networks and are educating citizens about climate change and its impacts. Others are working on developing a range of sustainability initiatives to not only adapt to climate change but also to lessen their contributions of greenhouse gases.

Despite the hard work of climate change deniers, the public is waking up to the fact that we are all living in a changing world. The evidence is everywhere. People are becoming far more concerned as they face the challenges associated with a warming planet. They are starting to make changes in their lives that seek to make improvements for themselves, their families, and their communities. The following section examines what we have done so far as a world to try to address climate change. As will be seen, the world has earnestly tried to advance a global climate change agenda. However, efforts have been partially thwarted by some countries that seek to advance their own agendas. As a result, the work of individual nations, states, and cities has gained in significance as a result of the inability of the world to come to consensus around climate change policy.

Examples of Impacts

There are many examples of places where climate change is having a direct impact. This section will review two locations: Kivalina, Alaska, and Manila, Philippines.

Kivalina is a small Innuit town in northwestern Alaska that rests on a low-lying barrier island in the Chuckchi Sea. Over the last two decades, it has undergone significant erosion and inundation due to the impacts of climate change (Shearer, 2011). Over the years, the sea ice, which normally protects the island during winter storms, has stopped forming for the length of the winter season. As a result, the shoreline of the island is unprotected and the island has been regularly eaten away by strong waves.

Kivalina decided to sue Exxon and other energy companies for damages in 2008. The lawsuit alleges that the oil companies produced products that led directly to climate change and that the village was undergoing severe stress as a result of climate change that required it to relocate. The suit sought $400 million in relocation costs. The lawsuit was eventually rejected by a US district court which decided that the problems associated with climate change were so big that they needed to be resolved by congress and the administration, not by the courts.

Of course, we know that the US Congress and administration have not resolved the issue of global climate change. Kivalina is expected to disappear as an island by 2025.

Manila, Philippines, is a sprawling city of about two million people which is facing serious threats due to climate change (Sengupta & Lee, 2020). Indeed, the IPCC has listed Manila and many other cities like Lagos, Shanghai, and Mumbai, as among the most vulnerable in the world to rising sea levels. What is particularly problematic about the situation in Manila is that there are tremendous numbers of people living in informal housing along low-lying areas that are frequently inundated during high tides and tropical storms. The informal housing is often made of weak construction material on stilts.

In addition, the changing climate will make Manila hotter and stormier—two conditions that will stress the population due to enhanced disease and stressed infrastructure such as electricity and water supply. What exacerbates the problem is that many areas of Manila are slowly subsiding due to groundwater withdrawals. The combination of subsidence and sea level rise has made some areas of Manila unlivable.

The problem is not just in Manila. The entire island nation of the Philippines is vulnerable to climate change and many coastal communities are facing existential problems. The president of the Philippines, Rodrigo Duterte, has been an advocate of climate change initiatives, but has also been highly critical of global climate change efforts for not accomplishing very much. Indeed, as was highlighted in the introduction to this chapter, the small island states of the Pacific have worked hard to try to forge some type of climate change agreement that would prevent them from becoming inundated in the coming decades. Unfortunately, larger economic interests have been winning out (as led by Australia) to prevent any clear agreement. As will be seen in the coming section, the world has come together to try to solve the climate crisis. The resulting agreement was modest and many major polluters are not participating. As a result, the individual actions of countries, states, and cities take on a more pressing significance.

What We Have Done So Far

There have been some very strong international attempts to try to address climate change by reducing greenhouse gases. The two most important of these are the Kyoto Protocol and the Paris Climate Accord. However, even with these agreements, greenhouse gases continue to rise in the atmosphere. Plus, as we will see, many countries have few limitations on greenhouse gas pollution while others have strict limitations—which makes the agreements rather unpopular in some nations. As a result of the many difficulties in developing and enforcing global climate change policy, local governments, businesses, and non-profits are leading important efforts to try to enact real substantive change.

International Efforts. The Kyoto Protocol was the first major climate change agreement that required binding agreements among nations (Maamoun, 2019). An earlier agreement, that was forged at the 1992 Earth Summit in Rio de Janeiro, called the United Nations Framework Convention on Climate Change, was a nonbinding agreement which sought to stabilize greenhouse gases in the atmosphere. The Kyoto Protocol, which emerged from this agreement, is a binding treaty. It was adopted in 1997 and came into force in 2005.

From the start, the agreement was controversial. It required very different things from different types of countries based on their development trajectory. Highly developed Western countries were required to significantly cut back on greenhouse gases, while lesser developed countries did not need to limit emissions. This meant that countries like India and China—both major greenhouse gas polluters—could continue to pollute, while other countries like the United States and Germany had to pull back their emissions.

Why was this approved? The basic argument is that countries like the United States and Germany have been on a development trajectory for about 200 years that released huge amounts of greenhouse gases into the environment. The use of this energy allowed them, and other countries like them, to create advanced economies and infrastructure. The decades-long cumulative use of fossil fuels led to high levels of greenhouse gases in our atmosphere. Countries like India and China have only started to utilize energy at the intensity of developing countries. To ask China and India, and countries like them, to curtail the use of fossil fuels, at a time when they are moving forward into a new era of industrial development, is unfair and would slow their progress. Plus, when per capita greenhouse gases are examined, China and India, and countries like them, have a very limited per person impact compared to the West.

The Kyoto Protocol was signed by most nations of the world. However, it was never ratified by the United States due to what many in the US Congress saw as its unfair approach. Canada withdrew outright from the Protocol. The European nations stayed in the agreement and have made significant advances in reducing greenhouse gases.

The Paris Climate Accord, which was approved in 2015, is an agreement that built on the Kyoto Protocol (Mahapatra & Ratha, 2016). According to the provisions of the Accord, each country is to develop a plan to address how they are going to reduce greenhouse gases and they are to share their plans with the world.

One of the major drawbacks of the plan is that it is not binding. Unlike the Kyoto Protocol, which is a legally binding treaty, the Paris Climate Accord does not put any legally binding restraints on countries that do not follow through with their agreements. As a result, the Accord is largely seen as a weak agreement. In addition, one of the largest emitters, the United States, has said that it intends to withdraw from the agreement at some points and plans to rejoin at others. This calls into question the effectiveness of the agreement at a time when we are starting to see major climate change impacts around the world.

In the past, the world has come together to solve other environmental problems. For example, the Montreal Protocol was an agreement that successfully addressed the loss of ozone in the ozonosphere (Ozone Secretariat, 2010). Chlorofluorocarbons, which were widely used in aerosols and as a coolant, were found to cause the destruction of ozone in the upper atmosphere. The Montreal Protocol, which was established in 1987, established international agreements to reduce the use and subsequent banning of the use of harmful chemicals. Unfortunately, the world has not been as successful in addressing the use of greenhouse gases.

While attempts have been made via the Kyoto Protocol and the Paris Climate Accord, they have not really stopped the problem.

Individual Governmental Efforts. While international climate agreements have fallen short, a variety of national, state, and city efforts have made important contributions to reducing greenhouse gases. Germany, for example, made a variety of important decisions to try to cut greenhouse gas emissions (Schwirplies, 2018). Since 1990, Germany cut greenhouse gas pollution by 36% and hopes to go even further and cut emissions by 55% by 2030. The nation did this by clear planning to reduce its dependence on fossil fuel energy production. It aggressively invested in green energy and decommissioned coal-burning power plants. Norway also made significant improvements. It focused heavily on electrifying the nation's vehicles. Today, half of all of the vehicles sold in Norway are electric.

Cities are also involved in advancing strong climate change policies. New York City was one of the first major global cities to develop a climate action plan to significantly reduce greenhouse gases (Graham et al., 2016). It focused heavily on energy efficiency of buildings, reducing emissions in the transportation sector, and green power generation. Today, New York City has one of the lowest greenhouse gas emissions per capita. In 2015, the per capita greenhouse gas emissions of New Yorkers were 6.1 metric tons of carbon dioxide equivalents compared with 19 metric tons for the United States as a whole.

The leadership for New York's sustainability initiative came from Mayor Michael Bloomberg who in 2007 worked to create a plan called PlaNYC. The plan focused on a variety of sustainability initiatives including parks, solid waste, brownfields, housing, waterways, transportation, energy, water supply, air quality, solid waste, and climate change. After Bloomberg left office, Mayor Bill De Blasio continued strong sustainability planning within a new plan called OneNYC which continues to focus on climate change while also focusing much more on De Blasio's interests in environmental justice. De Blasio has been a strong advocate for the Green New Deal and has committed the city to New York's version of the Green New Deal.

Some cities and communities have worked to benchmark their initiatives with other communities. Perhaps the best example of this is the

Florida Green Building Coalition's (FGBC's) green local government certification program (Florida Green Building Coalition, 2020). The FGBC uses a number of metrics to evaluate communities across the state of Florida as to their sustainability initiatives—including reduction of greenhouse gases. Many of the other areas of benchmarking within the green local government certification, such as the creation of green fleets and green park maintenance, contribute to reducing greenhouse gases.

At a much larger scale, an organization called ICLEI Local Governments for Sustainability works around the world to benchmark communities on greenhouse gas reductions and sustainability planning (ICLEI, 2020). The group emerged as a United Nations initiative in 1990 to provide guidance for local sustainability initiatives and is now totally independent of the UN. Today, nearly 2000 communities around the world are members (Fig. 2.2). What ICLEI provides to its members is access to knowledge and tools that help communities meet their sustainability goals. For example, they offer tools for creating greenhouse gas inventories and climate action plans.

For-Profit Organizations. Many business leaders have been strong advocates for climate action. There are a number of reasons for this. Some may be purely altruistic and recognize that it is the right thing to do for society as a whole. Others may be interested in having a strong climate change policy because it is on brand for their products. Still others may be interested in advancing climate change policy because they are concerned that their businesses may be impacted by climate change. Regardless of approach, many businesses have worked to advance a strong sustainability agenda in the last decade. In fact, most Fortune 500 companies have sustainability plans and many of them focus heavily on climate change. The sustainability plans set particular goals for the companies—including goals to reduce greenhouse gases.

Unilever, which is a large international firm which produces a range of consumer goods like food, beauty products, and personal care products, has a very detailed sustainability plan (Unilever, 2020). It focuses on nine main areas: waste and packaging, opportunities for women, sustainable sourcing, fairness in the workplace, water use, improving nutrition, health and hygiene, inclusive business, and greenhouse gases. Unilever sets a very ambitious goal for its greenhouse gas initiatives and seeks to

Fig. 2.2 Philadelphia, Pennsylvania, is a member of ICLEI which helps the city benchmark its sustainability initiatives with other cities around the world. Is your community a member?

obtain 100% of its energy used in manufacturing from renewable sources by 2030. It is also working hard to reduce greenhouse gases from its transportation needs and from its cooling. Many companies like Unilever have developed plans to significantly reduce greenhouse gases and have made commitments to meet the reductions suggested in the Paris Climate Accord in the absence of international consensus.

One of the more interesting developments as of late in the business world is the decision by BlackRock, the world's largest investment firm, to take a strong stand on sustainability and climate change. In January of 2020, the company's CEO, in a letter to shareholders, stated that the firm would no longer invest in companies that produced a high risk to the planet's sustainability—particularly those that contribute greenhouse gases to the environment (BlackRock, 2020). The risk, according to CEO Larry Fink, is just too great. The likely impacts from climate change in insurance, housing, and food markets are all detrimental and it is inappropriate to invest in companies that are damaging the planet. What this means is that BlackRock's considerable assets will no longer invest in fossil fuels companies like coal producers. Plus, the fact that BlackRock, the largest investment firm in the world, is embracing a sustainability agenda will have many ripple effects in the business world.

Climate Action Plans, Offsets, and Carbon Sequestration

There are two main ways that organizations can address greenhouse gas pollution to try to reduce greenhouse gases. They can develop a climate action plan or they can get involved with carbon sequestration initiatives.

A climate action plan involves the creation of the greenhouse gas inventory as well as a plan to reduce the greenhouse gases recorded in the inventory. Climate action plans typically involved a variety of organizational stakeholders. For example, if a company were doing a greenhouse as inventory, it would include representatives from all of the company's major departments that produce greenhouse gases such as transportation, manufacturing, and information technology. It may also include a range of company employees to add a range of knowledge to the plan. For example, a factory worker or a unit office manager may have more insight into how to reduce greenhouse gases than a CFO who is not involved directly in the frontline day-to-day operations of the company.

In conducting a greenhouse gas inventory, any organization should assess their Scope 1, Scope 2, and Scope 3 emissions as defined below:

Scope 1 emissions. These are the emissions that the organization produces through things like on-site power emissions or from vehicle use.
Scope 2 emissions. These are the emissions from purchased energy produced off site.

Scope 3 emissions. These are all other emissions that are associated with an organization. They can include emissions from employee commuting, wastewater, and landfill waste.

There are a number of tools available for organizations to conduct greenhouse gas inventories. As mentioned earlier, ICLEI for local governments provides a variety of tools and support for these initiatives. There is a great deal of data collection that must be done in order to complete a greenhouse gas inventory. This is, in part, why it is so important to have a representative group of stakeholders involved in completing a greenhouse gas assessment. Each unit of an organization is responsible in some way for producing greenhouse gases and thus should be involved in conducting the greenhouse gas inventory.

It is also important to calculate carbon offsets in the inventory process. Carbon offsets mitigate the greenhouse gas emissions of an organization. Perhaps the most obvious example is the production of green energy on site. For example, Tetra Pak's US headquarters in Denton, Texas, gets roughly 6% of its energy from an on-site solar array. However, it also purchases a considerable amount of green energy so that it obtains 100% of its energy use from renewables. What this means is that the overall Scope 1 emissions from electrical production are negligible. There are a variety of other ways to gain carbon offsets including the use of waste to produce energy and the enhancement of on-site carbon storage.

One of the most important ways that organizations are earning carbon offset credits is by purchasing them from organizations that manage off-site carbon mitigation projects. Companies and individuals can purchase carbon offsets for about $5 for 1000 pounds of carbon. Thus, if a company wanted to mitigate 100,000 pounds of carbon, they would pay $500 for those credits. There are many companies that provide certified carbon offsets. They take the money that they are given by individuals or companies to conduct the offsets and they invest the funds in projects that somehow reduce the carbon footprint of the organization. There is quite a diversity of carbon offset projects around the world. For example, some companies invest in large green energy infrastructure, others focus on restoring and preserving ecosystems, and still others focus on methane capture from landfills or other sites. Regardless of approach, the unifying theme in all of these projects is that they reduce the greenhouse gases of the organization.

Once an organization completes an inventory, the next step in the process is goal setting. Organizations must develop a range of temporal goals to cut greenhouse gas emissions. Often organizations set 25- or 30-year goals with 5-year mini-targets. For example, an inventory conducted in 2020 may set a goal of being totally carbon neutral by 2050. However, the organization may seek to reduce greenhouse gases by 10% by 2025 and set incremental reductions every five years. Regardless of the particular goal, the organization must set specific actions that need to be taken over time in order to reach the goals. The diversity of the team is important in this step. Greenhouse gas reduction can be challenging and change is often difficult. The more people involved with the climate action plan, the more successful it will be.

Once the plans are complete, the next step is starting to take action. Someone has to be in charge of implementing the plan and must work with the team in making the key decisions to reduce greenhouse gases. The team also must report back to the organization as to the progress of the initiative. Annual reports should be produced along with larger reevaluations along the five-year intervals. Many organizations have used the process of stakeholder engagement, greenhouse gas inventory, and goal setting to great success and have significantly reduced their greenhouse gases.

The reduction of carbon dioxide is important, but many are advocating for industrial-scale carbon sequestration to significantly reduce greenhouse gas concentrations in the atmosphere. Carbon sequestration occurs naturally on the planet in two main ways. First, plants can store large amounts of carbon in living and in dead plant material. That is why the loss of forest cover in places like Brazil and Indonesia, the decomposition of the tundra, and the burning of old plant materials stored in fossil fuels are so concerning. Second, we can store large amounts of carbon in geologic materials. There are a variety of ways that we do this, but perhaps the most obvious is in the formation of carbonate rocks. Carbonate rocks form from reefs and from fossilized carbonate animal shells, bones, and other materials.

In recent years, we have put the carbon cycle out of balance by altering the composition of the atmosphere and by burning carbon that has been stored by geological processes for millions of years. Many have developed methods for utilizing the carbon cycle to try to recapture carbon from the atmosphere and store it so that it is no longer a problem for planetary climate change. There are a range of techniques, but some of the most common options are the pumping of large amounts of carbon dioxide into deep underground storage. There are a number of geologic settings where carbon dioxide can be stored in deep reservoirs where it is nearly impossible for it to be released back into the environment. These places include old petroleum extraction sites as well as deep caves. The problem, however, is that vast amounts of carbon would need to be stored and it is currently very expensive to facilitate carbon storage. Nevertheless, a number of promising carbon storage prototypes have been built and some have even been made economically feasible.

The Remaining Challenge

We have done so much to try to address climate change but we have not solved the problem. Carbon dioxide levels continue to rise in the atmosphere and the prognosis for the future is not good. We are likely to continue to experience warming conditions in the foreseeable future. As a result, we will see major societal disruptions that will include shifting ecosystems and agricultural regimes. We are in the midst of a planetary crisis.

Yet, we can still solve the problem. The scientific community has very clearly outlined what we need to do. We need to rapidly decrease emissions of greenhouse gases. There are distinct political, social, and technological challenges that must be addressed.

- Challenge 1. Can we develop a global political consensus to solve climate change? To date, the progress on this challenge has been modest. We developed agreements like the Paris Climate Accord, but they are not aggressive enough to solve the problem. Plus, major greenhouse gas polluting nations like the United States, Brazil, and Australia have

been led by leaders who deny climate change. Can these countries develop the leadership they need to be part of the solution and not part of the problem?
- Challenge 2. Can we rapidly cut our reliance on fossil fuels and replace them with reliable green alternatives? Many nations have rapidly increased their production of renewable energy like wind and solar. Some countries like Germany now get the majority of their electrical production from renewable energy. However, renewable energy is a relatively small proportion of the global energy use and fossil fuels like petroleum, coal, and natural gas still dominate.
- Challenge 3. Can we transition to carbon-free future while maintaining a good standard of living in developed countries and while raising the standard of living in impoverished countries? The world, via the Millennium Development Goals, has made tremendous progress in a variety of development indicators. This has occurred while greenhouse gas emissions increased. Can we revise what it means to be developed as we transition off of fossil fuels?

Our ability to meet the above challenges will determine our future. The ability to come together politically, the ability to technically transition to green energy, and our ability to maintain a strong economy using clean energy are the keys to solving the global climate crisis.

The prognosis for whether or not we can meet this challenge is mixed. This chapter started by highlighting that Australia shot down an effort by leaders of the small low-lying Pacific island states to develop sound policy that would protect them from the rising seas. Plus, the policies of Presidents Bolsonaro (Brazil), Trump (United States), and others are harmful for our global climate. Yet there is hope. The world is much more aware of the problem and people are looking for solutions. There is an emerging political consensus that much more needs to be done. In addition, green technology, particularly around renewable energy, and energy efficiency is improving. Our business leaders are also starting to take a stronger stand to try to advance a sustainability agenda. Young activists like Greta Thunberg are pressuring the world's leaders to take much greater action to solve the climate problem. While there are challenges, there are also many solutions. There is reason to have hope.

The next chapter looks more broadly at the sustainability problems we face as a global society. It will review a number of issues within the three main themes of sustainability: environment, economics, and equity. As will be seen, climate change is but one of many complex issues we are facing on the planet. We live an interconnected global world and it will be shown that the choices that we make as individuals have far-reaching impacts.

References

Anderson, T. R., Hawkins, E., & Jones, P. D. (2016). CO_2, the greenhouse effect and global warming: From the pioneering work of Arrhenious and Callendar to today's Earth System Models. *Endeavor, 40*, 178–187.

BlackRock. (2020). A fundamental reshaping of finance. Retrieved December 5, 2020, from https://www.blackrock.com/corporate/investor-relations/larry-fink-ceo-letter

Boer, M. M., Resco de Dios, V., & Bradstock, R. A. (2020). Unprecedented burn area of Australian forest fires. *Nature Climate Change, 10*, 171–172.

Connell, J. (2016). Last days in the Carteret Islands? Climate change, livelihoods, and migration on coral atolls. *Asia Pacific Viewpoint, 55*, 3–5.

Constable, A. L. (2016). Climate change and migration in the Pacific: Options for Tuvalu and the Marshall Islands. *Regional Environmental Change, 17*, 1029–1038.

Cutter, S. L. (2006). *Hazards, vulnerability and environmental justice*. Earthscan.

EPA. (2019). Sources of greenhouse gas emissions. Retrieved June 1, 2019, from https://www.epa.gov/ghgemissions/sources-greenhouse-gas-emissions

Florida Green Building Coalition. (2020). Florida Green Building Coalition. Retrieved December 5, 2020, from https://floridagreenbuilding.org

Franta, B. (2018, September 19). Shell and Exxon's secret 1980s climate change warnings. *The Guardian*. Retrieved June 1, 2019, from https://www.theguardian.com/environment/climate-consensus-97-per-cent/2018/sep/19/shell-and-exxons-secret-1980s-climate-change-warnings

Global Monitoring Laboratory. (2020). Trends in atmospheric carbon dioxide. Retrieved December 5, 2020, from https://www.esrl.noaa.gov/gmd/ccgg/trends/

Graham, L., Debucquoy, W., & Anguelovski, I. (2016). The influence of urban development dynamics on community resilience practice in New York City after Superstorm Sandy: Experiences from the Lower East Side and the Rockaways. *Global Environmental Change, 40*, 1112–1124. https://doi.org/10.1016/j.gloenvcha.2016.07.001

Heede, R., & Oreskes, N. (2016). Potential emissions of CO_2 and methane from proved reserves of fossil fuels: An alternative analysis. *Global Environmental Change, 36*, 12–20.

Hoegh-Guldberg, O., Mumby, P. J., Hooten, A. J., Steneck, R. S., Greenfield, P., Gomez, E., Harvell, C. D., Sale, P. F., Edwards, A. J., Caldeira, K., Knowlton, N., Eakin, C. M., Iglesias-Prieto, R., Muthiga, N., Bradbury, R. H., Dubi, A., & Hatziolos, M. E. (2007). Coral reefs under rapid climate change and ocean acidification. *Science, 318*, 1737–1742.

ICLEI. (2020). ICLEI local government for sustainability. Retrieved December 5, 2020, from https://www.iclei.org

IPCC. (2014). *Climate change 2014: Synthesis report*. Retrieved June 1, 2019, from https://www.ipcc.ch/site/assets/uploads/2018/02/SYR_AR5_FINAL_full.pdf

IPCC. (2018). *Global warming of 1.5 °C*. Retrieved June 1, 2019, from https://www.ipcc.ch/site/assets/uploads/sites/2/2018/07/SR15_SPM_version_stand_alone_LR.pdf

Larson, E. (2020, March 17). Exxon loses jurisdiction fight in Massachusetts climate suit. *Bloomberg Green*.

Maamoun, N. (2019). The Kyoto protocol: Empirical evidence of a hidden success. *Journal of Environmental Economics and Management, 95*, 227–256.

Mahapatra, S. K., & Ratha, K. C. (2016). Paris climate accord: Miles to go. *Journal of International Development, 29*, 147–154.

McIntosh, E. (2019, June 14). Michael E Mann took climate change deniers to court. They apologized. *Grist*. Retrieved June 14, 2019, from https://grist.org/article/michael-e-mann-took-climate-change-deniers-to-court-they-apologized/

Ozone Secretariat. (2010). The Montreal Protocol. Retrieved June 1, 2019, from https://web.archive.org/web/20130420100237/http://ozone.unep.org/new_site/en/Treaties/treaties_decisions-hb.php?sec_id=5

Packham, C. (2019, August 13). New Zealand PM Calls on Australia to answer Pacific island climate change demands. *Reuters*. https://www.reuters.com/article/us-pacific-forum-australia/new-zealand-pm-calls-on-australia-to-answer-pacific-island-climate-change-demands-idUSKCN1V327B

Schwirplies, C. (2018). Citizen's acceptance of climate change adaptation and mitigation: A survey in China, Germany, and the U.S. *Ecological Economics, 145*, 308–322.

Sengupta, S., & Lee, C. W. (2020, February 13). A crisis right now: San Francisco and Manila face rising seas. *New York Times*.

Shearer, C. (2011). *Kivalina: A climate change story*. Haymarket Books.

Tucker, W. C. (2012). Deceitful tongues: Is climate change denial a crime? *Ecology Law Quarterly, 39*, 831–894.

Unilever. (2020). Sustainable living. Retrieved December 5, 2020, from https://www.unilever.com/sustainable-living/

3

Our Great Sustainability Challenge

Introduction

While climate change is a major existential threat to all of us, it is not the only major sustainability challenge we are confronting as a global society. Many areas of the world are facing serious challenges on a variety of fronts that call into question the long-term sustainability of major cities and regions. Yemen, for example, is facing a number of issues related to water, health, and security that are driven, in part, due to unsustainable use of its natural resources (Gladstone, 2019). At the same time, many areas such as Las Vegas, Nevada, survive only because of intensive, expensive, and difficult to maintain infrastructure. Las Vegas and many other parts of the world are avoiding Yemen-like difficulties only because they have the funds to expend to support their perilous situation.

We face a great sustainability challenge as a society. We are using natural resources far faster than they can be replaced and we are taxing the great ecosystems of the world beyond their ability to recover. We are changing the planetary systems in fundamental ways. As we change these systems, we also create even more unequal social and economic systems on our own communities, regions, nations, and the world. Unfortunately, the world is not a fair place. The data on infections and deaths from COVID-19 in the United States shows that poor communities and communities of color were most impacted by the pandemic across the United States and in individual communities like New York and Chicago. Similar issues emerge when one considers issues of environmental sustainability. Some of us are not significantly impacted by the unsustainability challenges that emerge across our planet and some of us are.

The information presented in this chapter can be depressing. It is never easy to learn about all the ways that we are harming our planet. Seeing a single impact in our community can be sad and learning about the cumulative impacts across the world can be heartbreaking. While this chapter focuses almost exclusively on the problems, much of the rest of the book looks at solutions. It is important to diagnose a problem before one can consider the appropriate cure. We are going to take a hard look at our planetary wounds.

This chapter focuses on some of the major sustainability challenges we are facing within three broad categories. The first category is environmental sustainability. This section will review a number of issues within the context of environmental sustainability such as water resources, pollution, energy, and food. The second category is economic sustainability. This section will focus on a variety of economic issues such as green economic development, the Green New Deal, and economic inequality. The third category is social sustainability and equity. This section will outline a range of issues such as environmental justice and environmental racism. As will be seen, there is overlap among these three broad categories and many of the issues transcend categorization. These three categories, known as the three E's (environment, equity, and economy), serve as the backbone for conceptualizing the modern sustainability movement. Prior to reviewing the categories, however, it is worth delving deeper into how we are changing our planetary systems.

Welcome to the Anthropocene

In the last decade or two, geologists have come to understand that our earth's systems are changing and that we are in a time of profound geologic change (Fig. 3.1). In fact, a new term for this time period has been suggested, the Anthropocene (Crutzen, 2006). There is some disagreement as to when the Anthropocene started. Some believe that it should start with the advent of the industrial revolution. Others suggest that it should be later and begin with the advent of the nuclear bomb. Geologically, the time difference does not matter that much since the earth is approximately 4.5 billion years old.

Regardless of when the Anthropocene started, geologists agree that our current era is seeing extremely rapid change. There are major alterations of basic earth systems underway. In the previous chapter, changes in the carbon cycle were discussed in some detail. However, this is not the only

Fig. 3.1 Nature is highly altered in the Anthropocene as a result of human activity. Signs of alteration can be found everywhere including in this image in France

cycle that is experiencing change. Nutrient cycles like the sulfur, nitrogen, and phosphorus cycles have all been altered. The earth's water cycle is also changing as the climate is changing. We are also seeing major species extinctions and losses of many ecosystems.

These changes are also altering the basic rock cycle. A great example of this is in the Louisiana delta region (Roberts, 1997). Due to dams and water control along the length of the Mississippi River, sediment which would normally enter the Gulf of Mexico is held back and stored upstream. As a result, there is a shortage of sediment making its way to the delta. This causes the existing delta sediments to erode as the delta itself subsides. The entire Mississippi delta region is shrinking due to human alteration of the Mississippi River drainage basin (Blum & Roberts, 2009). There are more obvious examples of human alteration of the rock cycle. In the last two decades, mountaintop removal mining in the Appalachian Mountains in the United States has significantly altered the landscape of the region. We also cause huge changes along coastal areas as we either build on or destroy coastal environments.

We are all trying to figure out how we live within the Anthropocene (Costanza et al., 2006). Some of us are already adapting to the changes by moving away from coastal areas and by changing our lifestyles. Others are seeking to become more resilient by fortifying their homes and communities and by seeking to live more sustainably. Still others are unable to adapt or become more resilient and will face the Anthropocene unprepared. As we see the range of sustainability challenges we are facing in the sections below, it will become clear that those who understand the scope of issues we are facing will be better prepared to take action and thrive under difficult conditions.

Environmental Sustainability

The first of the three E's, environmental sustainability, focuses on key themes within the traditional environmental canon. As was pointed out above, we have significantly altered many of the earth's systems due to the overconsumption of natural resources. Indeed, our natural resources are so stressed at this particular moment of our history that many regions of the world are facing existential problems. There are a number of topics that could be discussed within the context of environmental sustainability, but this section will focus on pollution, water quantity, habitat and endangered species, energy, natural resources, and waste. The purpose of these sections is to outline some of the problems we are facing. The solutions will be discussed elsewhere in this book.

Pollution. Pollution has been a driving force of environmentalism for decades. For example, the massive air pollution events that occurred in London and other European cities during the early years of the industrial revolution drove the development of a variety of environmental regulations that serve as some of the first examples of pollution management. While we have largely found ways to regulate and manage a variety of nuisance pollution problems, there are many issues that remain. It is impossible in these pages to review the range of modern pollution problems we face, but it is important to stress three main challenges: nutrient pollution of waterways, pollution assorted with a variety of mining processes, and plastic pollution of oceans.

Around the world, thanks to the technological advances of the green revolution, we have recognized that fertilization is the key to enhancing crop production. As a result, we have produced huge amounts of fertilizer in the last several decades (World Bank, 2019). Because crops cannot utilize all of the fertilizer that we add to our farm fields, large amounts runoff in surface water during rain events or when crops are irrigated (Chen et al., 2017). In addition, because meat production is so high across the planet—we have been growing meat production by double-digit percentage each decade as of late—animal wastes are also important sources of nutrients in the environment.

The main problematic nutrients are phosphorus and nitrogen. When they are present in excess in waterways, they can cause excess growth of algae and plants that would not normally grow in the system or grow in the quantity that the nutrient-rich environment now supports. When these plants die and decompose in the water, the decomposition process uses up oxygen, thereby depleting the water column of dissolved oxygen that fish rely on for survival. These conditions are called eutrophic conditions and the process described in this paragraph is called eutrophication.

Environments that become eutrophic are foul smelling and are frequently devoid of much animal life due to the poor environmental conditions.

Many of the world's lakes, rivers, and coastal environments are eutrophic. These places often have toxic algae blooms that cause a disruption in recreation and commercial activities. In Florida, for example, toxic algae blooms can disrupt commercial fishing and tourist activities. When systems are particularly impaired, dead zones can form. These are large areas that are devoid of most animal life. Dead zones have formed in coastal regions all over the world due to the runoff of nutrient water (Altieri & Gedan, 2015). One of the largest of these dead zones is located in the Gulf of Mexico near the Mississippi River delta. The Mississippi River is one of the largest drainage basins in the world. It is home to much of the agricultural activity in the United States and hundreds of sewage treatment plants release nutrient-rich effluent into its waters. Each year, the dead zones expand during the spring and summer as nutrient-rich snow melt mixes with water running off of fertilized farm fields and sewage effluent.

Nutrients can also enter our groundwater. Nitrate pollution of groundwater is common in some rural areas where runoff of nitrogen-rich waters enters local well systems. Drinking water with nitrates can impact the oxygen flow in blood leading to a number of health problems, particularly among the young. A condition known as blue baby syndrome, or methemoglobinemia, occurs under these conditions.

Another activity, mining, produces different types of problems. Illegal gold mining is taking place in many parts of the world, particularly in South America and Africa. While the mining itself produces serious problems, the real culprit in this process is the amalgamation of gold using mercury (Markham & Sangermano, 2018). Because this guerilla mining is unregulated, there are serious mercury pollution problems that are emerging in a number of locations. Once the mercury enters the environment, it is in small concentrations and cannot be seen with the naked eye. Many people around the world are exposed to harmful concentrations of mercury without knowing it.

There are a number of other forms of mining pollution around the world, but one of the most concerning is sediment pollution left over from the mining of tar sands. As the world has exploited the most easily accessible sources of petroleum in wells, we have moved to less accessible sources like tar sands which are sands that contain large amounts of hydrocarbons. Unfortunately, the tar-like petroleum cannot be pulled from the sand using traditional oil-extraction techniques. Instead, the tar sands must be mined and heated to pull out the petroleum. Sand and other materials are left behind as a waste product.

Places where tar sands are mined, such as in the Athabasca region of Alberta, are beset with many pollution problems (Lynch et al., 2016). The tar sands are mined using regional pit mining. After the oil is pulled out vast amount of waste sand is left on the landscape. Often, the sand contains trace amounts of oil and other contaminants and they become a vexing and unsightly long-term problem for the region.

Plastic pollution has emerged as a significant environmental problem in the last two decades (Morelle, 2019). Today, plastic has entered our food and drinking water systems. Plus, plastic pollution is a serious problem for our oceans, lakes, and rivers. Plastic is a very versatile material. It can be made into a variety of things and it has become one of the most used materials on our planet. Unfortunately, it does not break down easily in the environment and it has turned into a serious pollutant. There are a number of different types of plastic pollution issues, but two key ones are plastic pellets and plastic bottle and bag waste.

Plastic pellets are used in soaps, toothpastes, and beauty products as a gentle abrasive. While many nations now ban plastic pellets, they remain a lingering problem in many environmental systems (Dris et al., 2015). The pellets range in size and can be microscopic. As a result, plastic pellets are commonly found in fish and seafood.

Plastic pollution of bags and bottles (among other materials) has caused havoc with marine life. Most sea birds and turtles have ingested plastics and many marine and riverine animals have died from ingesting plastic. Plastic pollution is ubiquitous around the world. There are places where plastic is the dominant sediment and plastics clog riverways and coastal systems. Many parts of the world have moved to ban plastic bags and bottles, but the problem remains. We continue to produce plastics and the plastics we have thrown away will remain for generations.

While there are a variety of other pollution issues that could be reviewed, these three issues—nutrient, mining, and plastic pollution—serve as great examples of the range of problems we are facing. As we progress through our current century, it will be important to address these and other pollution challenges. Pollution can occur anywhere on our planet in our air, water, and soil. In areas where there are both pollution and water quantity issues, the sustainability challenges are multiplied.

Water Quantity. Water quantity is a growing problem in many areas of the world. Places like Athens, Tel Aviv, and Las Vegas, which are located in arid, semiarid, or Mediterranean climates, have distinct water challenges due to the lack of rainfall or the seasonality and irregularity of rainfall. They must pull water from great distances or they must develop new technologies, such as desalination, to supply the public with water.

Most water around the world, however, is used for agricultural and industrial purposes. While most water used for agricultural purposes does not need to be treated to drinking water standards, it can put a strain on a region's overall water budget. For example, in California, there is significant water policy conflict between thirsty cities like Los Angeles and San Francisco and the agricultural districts of the state's central valley.

The Aral Sea provides a great example of the types of damage we have done to water resources (Micklin, 2016). The Aral Sea was once the fourth largest lake in the world at nearly 70,000 km^2. Today, it is more or less gone. The shrinkage of the lake started in the 1960s as the region, now located between Kazakhstan and Uzbekistan, developed a range of agricultural initiatives that diverted water from tributaries of the lake for irrigation purposes. Today, the lake is less than 10% of its original size and the former lake bottom is now called the Aralkum Desert. While there are efforts underway to try to restore at least part of the lake, the reality is that humans, over the course of just a few decades, destroyed the fourth largest lake in the world.

There are many examples, although perhaps not as visually dramatic, of human destruction of water resources over the last few decades. Yemen is running out of water and the capital city of Sana'a is likely to face an existential crisis soon. The Ogallala Aquifer in the central United States is nearly tapped out. Plus, many coastal regions and island communities have withdrawn so much groundwater that saltwater has intruded into their aquifers making them unusable for centuries if not longer.

Groundwater withdrawal can also cause land subsidence and ecosystem disruption.

Habitat and Endangered Species. As noted earlier in the discussion about the Anthropocene, we live in a time when we are seeing massive extinction rates and loss of ecosystems (Burney & Flannery, 2005). Why are we causing these problems? One of the easiest culprits is population growth. In the early 1800s the world's population was about one billion people. Today, the population is nearly eight billion people and growing. We will most likely be over nine billion people this century. Our global population needs more and more resources, particularly as developing countries have improved their economies over the last few decades. We need more food, energy, housing, and a range of other materials that cause tremendous strains on planetary resources. As we seek, use, and dispose of resources, we put great strains on our planet's ecosystems.

But it is not just the number of people that is causing the problem. It is also HOW we live. Each person has an ecological footprint that impacts our planet. An ecological footprint is a measure of the amount of space we require to support our lifestyle. Some of us live very simply and have a low ecological footprint and some of us live lives that utilize lots of resources and thus have much higher ecological footprints. Wealthy countries like Luxembourg, Qatar, Australia, and the United States have very high ecological footprints. We can also calculate our personal footprints that tell us how we impact the planet. The consumptive lifestyles in the developed world significantly alter our planet in negative ways (ipbes, 2019).

One good example of how we are rapidly changing ecosystems due to our consumptive habits is the growth in the need for food—particularly animal protein. Across the planet, as developing countries have gotten wealthier, demand for meat has increased significantly—particularly in Asia. As a result of this, there is a need for greater meat production to satisfy demands. Unfortunately, the move from a plant-based diet to a diet that is more focused on meat has a much larger planetary impact. Agricultural fields that would normally produce vegetables for human consumption have to be transformed into crops that feed animals. Today, the vast majority of agricultural lands are utilized to support meat production in some way.

There are other agricultural products that have caused significant ecosystems disruptions in the last two decades. The growing demand for tropical palm oils has caused significant landscape transformation in Southeast Asia where many acres of tropical rainforest have been converted into palm oil plantations (Lam et al., 2009). In the last two decades many plants and animals have been lost to us forever due to its conversion.

The move to meat and products like palm oil has caused significant ecosystem disruptions as farmers seek to meet the global demand. The ongoing destruction of the Amazon is clearly linked to cattle production. Plus, vast areas of Africa are currently undergoing a transformation to industrial-scale agriculture in part to support the growing global meat demands. Of course, agriculture is not the only culprit in ecosystem and species loss. Suburbanization and our expanding cities gobble up lots of land as does a range of associated infrastructure like roads, ports, and airports.

The impact of all of this development on ecosystems is staggering. In the last few decades, there has been roughly a 60% decline in vertebrate populations and the rate of extinction is 100 to 10,000 times the rate that the world would normally face. Currently, approximately a million plant and animal species are on the verge of extinction with no end in sight.

One of the most diverse and productive ecosystems on the planet, wetlands, have been particularly impacted over the last century (Davidson, 2014). Wetlands are permanently, seasonally, or periodically inundated pieces of land that support a range of plant and animal species. Typical wetland types include swamps, marshes, tidal flats, bogs, and fens. Wetlands are sometimes seen as a negative component of the landscape because they are often odoriferous and insect-ridden and can hamper development. As a result, we have destroyed wetlands as we expanded agriculture and our cities (Li et al., 2014). According to the United Nations, 35% of the earth's wetlands were lost between 1970 and 2010 and that rate is accelerating. In the United States, more than half of its wetlands have been drained. One of the major reasons the world is losing coastal wetlands is because of climate change driven by our energy use.

One of the most troubling issues with habitat loss is extinction of plant and animal species. Several well-publicized extinctions occurred in the last few decades (Kretzschmar et al., 2016). For example, the Pyrenean ibex went extinct in 2000 and there are only a handful of western black rhinos and northern white rhinos left in the world.

Energy. Our quest for energy is one reason we have seen so much land conversion in recent years. Overall, we get most of our energy (roughly 85%) from the main hydrocarbon sources: oil, natural gas, and coal. We only get 7% of our energy from hydroelectric power and 4% from nuclear power. All of the renewable sources total only 4% of our energy budget. Thus, while there has been great movement on renewables, the reality is that we still rely on fossil fuels for the vast majority of our energy use.

Energy production has some serious sustainability challenges. The extraction of oil is inherently messy. The Deepwater Horizon oil spill in the Gulf of Mexico that began in April 2010 leaked over 200 million gallons (Reddy et al., 2012). The famous Exxon Valdez oil spill leaked over 10 million gallons off of coastal Alaska in 1989. Many other less famous spills have occurred in recent decades. Each oil spill causes major environmental disruptions such as pollution of water, damage to ecosystems, and respiratory problems for humans. While many safety mechanisms have been put in place to try to prevent the worst outcomes from these types of events, they still do occur.

Hydraulic fracking, or more simply, fracking, is another problematic energy extraction process. Fracking is a process that seeks to extract natural gas or petroleum that is tightly held in geological materials that could not be extracted by normal pumping processes. Holes are drilled into the rock and fracking fluids are forced under high pressure to break up the rock to release the natural gas or petroleum which is then pumped out of the subsurface.

The problem with fracking is that there is great public concern over the potential for groundwater contamination. Fracking fluids mainly contain water and a little bit of sand along with unknown chemicals (Pierre-Louis, 2017). In the United States and many other parts of the world, fracking companies do not need to disclose the composition of the unknown chemicals because they are exempt by federal law from reporting their contents. Fracking is usually done in rural areas where there is a strong reliance on individual wells that are not normally tested for water quality. The mystery around the fracking fluid is concerning to many.

People want to know what is being put into the ground in order to understand if their drinking water is safe for consumption.

Natural Resources. The use of natural resources overall is another important theme of environmental sustainability. We are using natural resources like aluminum, rare earth metals, and even helium at rates that far exceed the earth's ability to produce them. For example, rare earth metals, used in cell phones and other electronics are exceedingly rare. They are found in minable concentrations in only a few corners of the world. Our electronic technology is very popular and thus the demand for rare earth minerals is leading to greater mining of the materials. Because the minerals are extremely rare, the amount of land needed for the mines is extensive and the wastes associated with rare earth mining are of great volume.

Helium is another natural resource that we are losing. It has many practical applications although it is currently being squandered for use in party balloons. Although helium is a component of the atmosphere, and is actually relatively abundant, it is very difficult to find in concentrations that are useful for commercial application. The problem with helium is that it is literally lighter than the other gases in the atmosphere and can leave out atmosphere when it is released into the air. Most commercial helium is mined in underground reserves where it is associated with fossil fuels. These reserves have diminished in recent decades and the cost of helium has jumped.

The mining of other natural resources has other problems. Diamonds, for example, are gemstones that are highly prized due to their cultural significance (Fig. 3.2). They are given as symbols of love and are often used as status symbols. For example, in some cultures, it is thought that an engagement ring should cost the equivalent of three months of salary of the man or woman who purchases it. The quest for diamonds over the years created a number of problems. For example, the Apartheid system in South Africa created great inequality between mine workers and mine owners and managers. Conditions for the workers were akin to slave-like conditions as workers were often separated from their families for long periods of time. The high value of diamonds and other gemstones has led to a rush of illegal mining around the world with subsequent pollution and social disruption problems.

Fig. 3.2 Our quest for more and more luxury goods has a direct impact on the environment

Unfortunately, because we live in such a consumerist society, we have major natural resources problems because we throw so much stuff away. We do not repair or reuse things as much as previous generations. We are always in the quest for the new. Nothing exemplifies this current trend more than our thirst for fast fashion. Typically, fashion seasons followed the annual season: fall, winter, spring, and summer. Today, however, fashion companies like H&M are constantly churning out new looks and new trends. We are buying twice as many clothes as we did just a couple of decades ago, and more and more clothing is entering the waste stream as we cycle old looks out of our closets.

The same problem exists for home furnishings and décor. Television shows on networks like HGTV encourage the new and improved. They set a tone that suggests that we need to constantly upgrade, buy what is in fashion, and redo the old. They have helped to turn homes and home décor into a form of fast fashion. The constant drive for the new creates many problems for waste managers who have to deal with the range of stuff left behind in the world's constant drive for consumer good.

Waste. Around the world waste production is up. This is not a surprise since as the world has gotten wealthier, it is consuming more short-term consumer goods. The World Bank estimates that world waste production could increase by 70% if no action is taken (World Bank, 2020). There are a variety of ways that we manage our waste:

- Sanitary landfills. These systems are essentially highly technical dumps. They are often lined with impermeable materials so that liquids that filter through the landfills are not released into the subsurface. The liquids, called leachate, often contain hazardous chemicals that it picks up as it moves through the debris. Sanitary landfills are expensive to build and maintain. Some of them are highly specialized and can handle hazardous waste.
- Unsanitary landfills. These are essentially open dumps that are not managed for environmental purposes. They usually are unlined or have limited filtering capacity. As a result, the leachate can enter groundwater systems near these dumps.
- Waste to energy. These systems are among the fastest growing ways that we manage waste around the world. Waste is collected and taken to a power generation facility where the waste is burned as fuel. The actual energy production is not all that different from a coal or natural gas power plant. The energy produced by waste produces steam which turns turbines to create electricity. The newer waste to energy power plants produces limited emissions. However, the remaining ash from burning the waste can contain heavy metals and be difficult to dispose.
- Open burning. There are some parts of the world that allow open burning of waste due to a lack of regulation. In such situations, emissions from burning waste can be toxic and cause problems for surrounding communities.
- Ocean dumping. This type of waste management obviously causes serious problems for ocean ecosystems. Unfortunately, this is a common type of waste management in some areas of the world.

The best way to deal with waste is clearly not to produce it. In the last several decades, communities around the world have campaigned to cut waste using the three R's of reduce, reuse, and recycle. In some areas of the world, these campaigns have been successful. Globally, however, we continue to increase our waste production.

Recycling waste is also problematic. Much of the West's waste is carted off to the developing world for handling. Unfortunately, some of this waste upon arrival never makes it into the recycling stream and it ends up being a lingering waste problem for the country of import (Watson et al., 2019). As a result, some countries have stopped accepting imported plastic and electronic waste and there are questions as to the viability of recycling of these common materials. Even the vast majority of used clothing ends up going to the waste stream. We just produce too much clothing for the used clothing market and there are limited uses for the waste fiber. Food waste is also up. Food waste makes up anywhere from 10% to 20% of the volume of municipal garbage. Food waste in the United States accounts for about 40% of all the food that is produced. There are a number of reasons for this. In some cases, food waste is caused by purchasing too much food. It is also caused by disposing of food that does not make it to market in time. There are also well-publicized cases of restaurants throwing out food that is not sold at the end of the day. Regardless of the cause, the world has a serious food waste issue.

Economic Approaches to Sustainability

One of the great challenges in the modern sustainability movement is how to advance a sustainability agenda while also creating strong economies. Some have argued that sustainability and modern economic systems are incompatible while others have said we need to work with current economic systems to enact structural changes that will help our environment. It is worth comparing and contrasting these ideas in order to better understand the economic E of the three E's of sustainability.

The Sustainability Critique of Capitalism. One of the most important critiques of the current world system is that capitalism has led to a fundamentally unsustainable and unequal world (Klein, 2010). In this argument, capitalists work against the greater good for individual gain. Perhaps the best example of this is the role that the major energy companies played in seeking to deceive the public around climate change. Some worked hard to publicly deny climate change even though their own scientists were sounding the alarm. Many question whether or not real progress on climate change can be made when governments in places like the United States, Russia, Venezuela, and Canada support their carbon producers with strong government policies.

It is important to stress, however, that the very nature of neoliberal capitalism present in most of the world creates serious challenges for sustainability. Neoliberalism is an economic philosophy that promotes free trade, deregulation, and privatization. Over the last several decades, neoliberalism has advanced around the world as globalization initiatives like the North American Free Trade Act or even the European Union promoted economic growth. Neoliberalism eschews regulation, including environmental and sustainability governmental policies. In addition, neoliberalism promotes the use of natural resources to advance economic development without check. Naomi Klein, in her book *Shock Doctrine*, notes that unpopular neoliberal policies often are thrust on populations in times of extreme social stress due to war, natural disasters, and terrorism. These policies, whether in Pinochet's Chile, post-tsunami Sri Lanka, or post-9/11 United States, benefit organizations and corporations with the greatest access to government.

The shock doctrine was recently on display in the United States as the federal government sought to stabilize the economy due to the quarantines needed to slow the spread of COVID-19. While there was some limited relief given to citizens, the real bailouts focused on large corporations. Instead of developing universal healthcare or universal COVID-19 testing, the government focused on bailouts of cruise ship companies, airlines, and a variety of businesses. Under even the most desperate circumstances, the focus of the government was not on helping the individual. It was on helping corporations. Given this tableau, it is not particularly surprising that neoliberal governments are not especially effective at enacting viable sustainability initiatives.

Another important critique of modern capitalism and its role in sustainability is that sustainability initiatives cover up bad practices. Walmart, for example, is well known for its work on developing sound sustainability practices within its supply chain and for making its stores highly energy efficient. However, the business model of Walmart is to sell as much stuff for as cheaply as possible to as many people as possible. Walmart is arguably responsible for the loss of many small businesses and has driven our modern culture of overconsumption. While Walmart (and many other companies) does a great deal to advance real sustainability, the heart of their business goes against sustainability tenets.

Capitalism in Support of Sustainability. There are many initiatives whereby capitalism has advanced a strong sustainability agenda. For example, green economic development is a tool that has been applied to transform local and regional economies (Brinkmann, 2018). Green economic development is typically managed by public-private partnerships that bring together a range of stakeholders involved with promoting the economic and sustainability interests of the community. For example, a community may come together to advance renewable energy production in their region and amass the forces of leaders in education, business, labor, and sustainability to encourage economic development around renewables.

Associations of businesses may also seek to advance sustainability within their collective enterprise. The Carton Council of North America, for example, formed to try to advance recycling of food and beverage cartons in North America. There are many examples of these types of industrial initiatives. There are also many third-party organizations that also measure, assess, and certify business activities and products. Energy Star, for example, will verify the energy use of appliances and other products.

Perhaps the most successful green capitalist venture in recent years is the movement to organic food. The production of organic food has doubled in the last decade in the United States and the number of certified organic farms is also increasing rapidly. The success in organic farm can be attributed in part to the public's growing interest in clean and healthy food. However, organic farming has been embraced by agribusinesses that recognize the growing demand. As a result, organic food has moved mainstream in the last decade.

While the success of organic food is to be lauded, it must be noted that it remains a small component of the overall food market. Most food produced in the United States remains tied to conventional farming using the tenets of the green revolution. Thus, like many aspects of sustainability, there are strong initiatives, but they have not fully transformed practices to the point that the sustainability approach dominates the market.

This is the challenging aspect of the capitalist approach to sustainability: while some solutions are viable and are being used by some, the overall culture has not widely adapted sustainability solutions. We see this not only in organic food, but in many areas such as electric car use, organic clothing, consumption of consumer goods, and renewable energy. Within the capitalist system, some may choose to go green, but many do not. They prefer to live their lives for the greater good of the individual or the business, not the community as a whole. We see this played out not only with individuals and businesses, but also with local, national, and regional governments. Decisions are made to cut taxes and services or to support short-term interests without taking into consideration the long-term impacts of government actions on society, economies, or the environment. The waffling between short-term and long-term visions for the future defines the recent histories of many national governments such as Brazil, the United States, and the United Kingdom. Leaders with long-term visions for a sustainable future like Barack Obama contrast sharply with those of Donald Trump.

Transformed Business Models. Several decades ago, in 1973, E. F. Schumacher published a book called *Small Is Beautiful: Capitalism as if People Mattered* that critiqued the modern capitalist system as too big and too heartless (Schumacher, 1973). He noted that it led to income inequality, unfairness, and environmental degradation. He believed that we needed to transition to much smaller forms of economic development focused on small businesses that had smaller environmental footprints. This book influenced many business-minded people all over the world such as Ray Anderson, the founder of Interface Carpeting, one of the largest carpeting manufacturers in the world. After reading Schumacher's book, he realized that his company was harmful to the planet and he needed to change it. He made a decision to get rid of harmful chemicals, focus on reuse and recycling of carpeting, and utilize natural materials whenever possible. He also sought to become a carbon neutral company. He transformed Interface into one of the greenest companies in the world (Anderson & White, 2009).

Businesses like Interface transformed the three E's of sustainability—equity, environment, and economy—into the three P's, people, planet, and profit. These organizations developed sustainability plans, conducted greenhouse gas inventories, and found ways to transform their business models so that their activities are not as harmful as they once were. Most Fortune 500 companies now have sustainability plans and hire sustainability professionals to guide them in how to become more sustainable organizations. In many ways, these companies are leading the way in global sustainability in the absence of strong governmental leadership.

It can be argued that even though many of these companies have transformed themselves, they are still responsible for global unsustainability. For example, Interface may be a green company, but their carpeting is used in airports, hotels, and a variety of settings that may not be sustainable operations. Walmart may have transformed their supply chain but their business model promotes unsustainable consumption. Because of these issues, some have taken deeper steps to promote green enterprise within Schumacher's vision of an economy as if people and the planet mattered.

Sustainabillies and Minimalists. I coined the term sustainabillies to refer to people who have rejected the status quo in society and formed their own green lifestyle and way of business. They have left the consumption rat race and do not need or want as much stuff. Some of these people may call themselves minimalists and not necessarily identify with a sustainability agenda. Regardless, there are growing numbers of people around the world who are making distinctly different lifestyle choices including how they earn a living. They may live in small homes or cohousing, live without cars, and/or reject traditional career paths.

People who have embraced this approach may have small businesses that focus on sustainability in some way. They may be farmers or food producers, craftspeople, or gig workers. They may opt to work less hours to focus on their own personal interests. They intentionally chose to not engage with the mainstream economic system and instead build support networks of mutual aid. There are many examples of people who have made these types of personal transitions. While we all have some type of impact on our planet, people who have embraced this approach to living have a far smaller environmental footprint than most of us.

Not all of us can make, or want to make, this type of commitment to environmental sustainability. Therefore, many will work within existing economic systems to try to make them as green as possible. As will be seen in the upcoming section, there are many equity issues within the realm of sustainability that need attention as well.

Equity and Sustainability

The environmental movement that swept the world in the 1960s and 1970s is often critiqued as largely a white, middle-class movement that did not include the concerns of the poor, the global south, or people of color. As Chris Sellers points out in his book, *Crabgrass Crucible*, a great deal of the activism of the era emerged in the suburbs out of concerns suburban communities had about the environmental contamination they were seeing in their areas (Sellers, 2012). Certainly, the environmental regulations that emerged from this activism benefited all of us. But because of the origins more or less emerged from the suburbs, many people were not represented in the eventual decisions that led to much of the environmental policy we have. As a result, a range of environmental equity, justice, and racism problems emerged in the last several decades.

Environmental equity or justice is the idea that all people should share the benefits and burdens of environmental issues. For example, when considering environmental benefits, all people should have access to environmental amenities like parks. They should not exist just for the elites amongst us. In addition, environmental burdens, for example, exposure to pollution, should not be the burden of one group of people. Environmental racism occurs when one racial group experiences the environmental burdens at disproportionate rates.

Environmental Justice in the United States. One of the first individuals to identify problems with environmental justice in the world was Robert Bullard who wrote many books on the topic including *Dumping in Dixie* (Bullard, 2000). In the book, he highlights how hazardous waste dumps are located disproportionally in African American communities in the American South. As a result, these communities experienced the unequal environmental burdens due to the presence of these facilities. While not every dump can be cited as evidence of environmental racism, the overall trend highlights the problem of institutionalized environmental racism.

As a result of this work and others, the Environmental Protection Agency (EPA) has an office of environmental justice that seeks to help communities work through environmental justice problems. Indeed, by shedding light on the issue of environmental justice, Bullard has helped to transform how politicians, businesspeople, and community members think about sustainability in their communities. While there are definitely problems, people are much more aware today about environmental justice and real efforts are made to avoid environmental justice problems before they happen.

Bullard's work began a new wave of research on equity within sustainability. The results demonstrated that there are many environmental justice issues not just in the United States, but around the world. Plus, while people are more sensitive to the problem, environmental justice issues have not disappeared.

Perhaps the most important and recent environmental justice issue that occurred in the United States was the Flint, Michigan, water disaster (Katner et al., 2018). Flint, Michigan, is a rustbelt community that in 2014 changed its drinking water supply from that provided by Detroit to a local source, the Flint River. Unfortunately, the water managers did not treat the water properly and it entered the water supply as a corrosive agent that released lead from pipes into the drinking water of the users. In addition, the changeover resulted in a spate of Legionnaires disease that caused deaths and many illnesses. The issue is complex and cannot be fully covered in this section. However, a range of bad decision-making led to this problem. The Flint disaster is widely seen as an environmental justice issue because the decision-making was largely made by white managers and the impacts were largely felt in minority communities.

Global Environmental Justice Issues. There are a range of global environmental justice issues that have emerged in the last few decades. These include indigenous rights, movement of garbage, and globalization of industries. What is common about each of the examples that will be briefly highlighted is that they all represent the imposition of negative environmental burdens on communities that have limited power to prevent the impacts.

One of the areas of concern is the impact of environmental problems on indigenous communities. An important example of this is the recent fight in the United States over the Dakota Access Pipeline (Brinkmann, 2017). This pipeline was designed to bring petroleum from the oil fields of Canada and the northern United States to the refineries to the south. The pipeline abuts the Native American reservation of the Standing Rock Sioux and there have been many protests in Native American communities to try to prevent the pipeline. The concern is that if there are leaks in the pipeline the oil would contaminate groundwater and surface water that comes into the reservations. Specifically, the pipeline crosses the Missouri River only half a mile from the reservation border. The Standing Rock Sioux rely on this river as their main source of drinking water.

Although protests and lawsuits went forward, the US government decided to move ahead with constructing the pipeline in 2017 and it is now active. However, the protests galvanized Native American communities and their supporters largely due to the strong reaction of federal officials who used pepper spray and dogs to try to disperse protesters.

This is but one example of indigenous communities under attack due to the forces of economic development. The recent land grab by Brazilian farmers and loggers in the Amazon rainforest is also causing significant problems for indigenous communities (Brinkmann, 2019). Many indigenous leaders who have tried to protect their lands have been murdered as encroachment advanced. This land grab is supported by Brazil's president and congress. As a result, indigenous people in Brazil have very limited options to try to preserve not just their lands, but their culture. Previous governments in Brazil developed policies to protect the rainforests and the native people who lived within them. Now, however, those rights are being taken away and the forests and the people are being lost.

The international transportation of garbage and recyclable materials is another example of an international environmental justice issue. Perhaps one of the more challenging issues is the recycling of plastics and electronic waste. Since their advent in the middle of the twentieth century, plastics have been used in a myriad of products. Indeed, plastic is an incredibly versatile material that can be molded into a range of products like dishware and eating utensils, food containers like plastic bottles, and automobile components. Plastic comes in a range of chemical compositions. It can be hard or soft, clear or colored, and thick or thin. Unfortunately, plastic does not readily break down and it remains a solid material long after it is created. As a result, it causes unique problems for the waste and recycling stream—in part due to its chemical diversity.

Many communities around the world have developed recycling programs that mandate that residents and businesses separate out plastic from the rest of the waste for recycling. The plastic is then collected by waste handlers and sent to a waste transfer station where the plastic is baled and shipped to a plastic recycling facility where the plastic is turned into new products. Some communities have single stream recycling which means that all potential recyclable material (paper, metals, glass, and plastic) are collected at once. The recyclables are then sent to a sorting center where the waste goes through a complex series of mechanical and robotic sorting followed by a final round of human sorting. The plastic that is separated from the rest of the waste is again baled and shipped out for recycling.

There are three main problems with this process. The first one is that the high variability of plastics in a recycling stream makes the job of sorting incredibly difficult. Different plastic types and different plastic colors need different recycling processes. Plastic bales are often heterogenous in plastic type and thus less valuable in the market. Indeed, if a plastic bale is highly contaminated, some recycling facilities cannot use the plastic. In such cases, all the work of sorting the plastic from the main waste stream is lost and the baled plastic becomes a new waste problem. The second problem is that there is not a giant market for plastics. While some products are made from plastics, such as plastic furniture, fibers, and deck planks, the reality is that recycled plastics are not highly desired in the open market. As a result, the value gained by communities for selling their recycled materials is very low. It costs more to recycle plastic than communities get from seeling the materials in the market. The third problem is that plastic recycling is not a lucrative business. As a result, millions of pounds of plastics are shipped from developed countries to developing countries with cheap labor and poor environmental regulations.

Over 70% of the plastic waste shipped from the United States to countries like China, Thailand, and the Philippines is processed in facilities with poor waste management systems. What this means is that relatively large percentages of the plastic that is shipped end up being disposed of in the ocean or poorly maintained landfills, or the waste is resold for local use. The local users may melt the plastic in unsanitary conditions leading to air pollution problems, or they may actually use the plastic as a cooking or heating fuel, again, leading to issues of air pollution.

The reason that the plastic issue is an environmental justice issue is that Western countries are exporting their waste to developing countries where the plastics end up causing a range of pollution problems from litter and ocean pollution to air pollution from the burning of the waste. Western countries spend a tremendous amount of money and time sorting plastics in their homes and in recycling facilities only to have the waste end up overseas where it harms local communities.

The exact same situation exists for electronics recycling. It is far cheaper to ship electronic waste overseas for processing. However, only certain components of electronics are valuable and much of the electronic waste ends up entering the waste stream of the country of import. Electronic waste is much more problematic than plastic waste because it contains so many different components which can lead to much more harmful pollution problems.

Because of local outcry in the communities that receive plastic pollution from overseas, many countries are banning the import of plastics from developed countries. While the problem still exists, there are fewer opportunities to ship plastics overseas and developing countries are struggling with how to manage their recycled plastics. In the United States, more plastic recycling operations are opening up. Yet around the world, plastic waste remains a vexing environmental justice issue.

International globalization and neoliberal policies, of course, are partly responsible for the environmental justice problems of plastics. It is easy to import and export plastic waste due to liberal trade policies. Indeed, the globalization of industry has emerged as one of the biggest environmental justice issues the world is facing. Global industries with headquarters in developed countries are taking advantage of cheap labor and lax environmental regulations in developing countries.

There are many examples that could be cited about the negative impacts of globalization of industries on the environment and on society as a whole. The plastic recycling industry is just the tip of the iceberg. Perhaps the best example of poor practices comes from the clothing manufacturing industry where workers toil under very difficult condition and where dyes and other materials are released into the environment without treatment.

In 2013, for example, a clothing factory in Dhaka, Bangladesh, that made products for companies like Gucci, Benetton, Versace, and Prada, collapsed killing over 1100 people and around 2500 were injured (Karim, 2014). Owners of the factory knew that the building was dangerous and still required their workers to show up or be fired. This example shows the direct impacts of consumerist societies of the developed world on societies in the developing world. Due to the lax regulation in areas of textile production in Bangladesh and other countries, rivers are widely polluted. Globalization leads to exploitation of workers and the export of pollution (Miller, 2012). Many countries are as clean as they are not from stopping their polluting lifestyles. They are clean because they are exporting their polluting lifestyles.

Environmental justice is emerging as one of the most important aspects of the modern sustainability movement because it brings together issues of economy and environment to solve problems that matter directly to communities. Many of the major environmental issues we face have environmental justice themes. For example, climate change is clearly an environmental justice issue when examined from the viewpoint of the small island states. They are relatively powerless to try to solve their existential issues when major economic powers like Australia and the United States are actively seeking to advance their carbon economies.

Summary

This chapter highlighted a number of sustainability problems we are facing around the world within the three major themes of sustainability: environment, economy, and equity. The environmental theme emerged from the traditional environmental movement and tends to focus on issues like pollution, ecosystem protection, and energy. Due to the vast changes the world is seeing in the Anthropocene, many of the earth's natural environmental systems are stressed and much work is being done to try to repair them. Economic aspects of sustainability follow two broad themes. The first theme focuses on working within existing economic systems to try to lessen their impact on the environment. For example, many businesses like Walmart have developed sustainability plans and many companies are seeking to benchmark their actions to assure their stakeholders that they are doing all that they can to act sustainably. The second theme is highly critical of neoliberal economic policies that have emerged in the last several decades and are seeking to transform existing economic systems so that they are not so hard on environment and society.

The equity aspect of sustainability has emerged most recently as one of the more important themes of the field. First described in the United States within the context of the citing of hazardous waste dumps, environmental racism remains a potent factor as was highlighted in the water quality problems that emerged in Flint, Michigan. Environmental justice issues also exist internationally, particularly among indigenous communities. The forces of globalization have also led to significant inequalities around the world in terms of labor and environmental quality. As we will see in the next chapter, there are serious ethical dimensions of sustainability that must be considered in order to evaluate our global sustainability progress.

References

Altieri, A. H., & Gedan, K. B. (2015). Climate change and dead zones. *Global Change Biology, 21*, 1395–1406.

Anderson, R. C., & White, R. (2009). *Confessions of a radical industrialist: Profits, people, purpose—Doing business by respecting the earth*. St. Martin's Press.

Blum, M. D., & Roberts, H. H. (2009). Drowning of the Mississippi Delta due to insufficient sediment supply and global sea-level rise. *Nature Geoscience, 2*, 488–491.

Brinkmann, R. (2017, February 26). Dakota Access Pipeline teaching resources. *On the Brink*. https://bobbrinkmann.blogspot.com/2017/02/dakota-access-pipeline-teaching.html

Brinkmann, R. (2018). Economic development and sustainability: A case study from Long Island, New York. In R. Brinkmann & S. J. Garren (Eds.), *The*

Palgrave handbook of sustainability: Case studies and practical solutions (pp. 433–450). Palgrave Macmillan.

Brinkmann, R. (2019, November 19). Brazil deforestation continues and indigenous people fight for survival amid calls for divestment. *On the Brink.* https://bobbrinkmann.blogspot.com/2019/11/brazil-deforestation-continues.html

Bullard, R. D. (2000). *Dumping in dixie: Race, class, and environmental quality* (3rd ed.). Westview Press.

Burney, D. A., & Flannery, T. F. (2005). Fifty millennia of catastrophic extinctions after human contact. *Trends in Ecology & Evolution, 20,* 395–401. https://doi.org/10.1016/j.tree.2005.04.022

Chen, J., Qian, H., & Wu, H. (2017). Nitrogen contamination in groundwater in an agricultural region along the New Silk Road, northwest China: Distribution and factors controlling its fate. *Environmental Science and Pollution Research, 24,* 13154–13167.

Costanza, R., Mitsch, W. J., & Day, J. W. (2006). A new vision for New Orleans and the Mississippi delta: Applying ecological economics and ecological engineering. *Frontiers in Ecology and the Environment, 4,* 465–472.

Crutzen, P. J. (2006). The "Anthropocene". In E. Ehlers & T. Krafft (Eds.), *Earth system science in the Anthropocene* (pp. 13–18). Springer.

Davidson, N. C. (2014). How much wetland has the world lost? Long-term and recent trends in global wetland area. *Marine and Freshwater Research, 65,* 934–941.

Dris, R., Imhof, H., Sanchez, W., Gasperi, J., Galgani, F., Tassin, B., & Laforsch, C. (2015). Beyond the ocean: Contamination of freshwater ecosystems with (micro-) plastic particles. *Environmental Chemistry, 12,* 539–550.

Gladstone, R. (2019, March 27). Cholera, lurking symptom of Yemen's War, appears to make roaring comeback. *The New York Times.* Retrieved June 1, 2019, from https://www.nytimes.com/2019/03/27/world/middleeast/cholera-yemen.html

ipbes. (2019). Media release: Nature's dangerous decline 'unprecedented'; species extinction rates 'accelerating'. Retrieved June 1, 2019, from https://www.ipbes.net/news/Media-Release-Global-Assessment

Karim, L. (2014). Disposable bodies: Garment factory catastrophe and feminist practices in Bangladesh. *Anthropology Now, 6,* 52–63.

Katner, A. L., Pieper, K., Lambrinidou, Y., Brown, K., Subra, W., & Edwards, M. (2018). America's path to drinking water infrastructure inequality and environmental injustice: The case of Flint, Michigan. In R. Brinkmann &

S. J. Garren (Eds.), *The Palgrave handbook of sustainability: Case studies and practical solutions* (pp. 79–97). Palgrave Macmillan.

Klein, N. (2010). *The shock doctrine: The rise of disaster capitalism*. Picador.

Kretzschmar, P., Kramer-Schadt, S., Ambu, B., Bender, J., Bohm, T., Ernsing, M., Göritz, F., Hermes, R., Payne, J., Schaffer, N., Thayaparan, S. T., Zainal, Z. Z., Hildebrandt, T. B., & Hofer, H. (2016). The catastrophic decline of the Sumatran rhino (Dicerorhinus sumatrensis harrissoni) in Sabah: Historic exploitation, reduced female reproductive performance and population variability. *Global Ecology and Conservation, 6*, 257–275.

Lam, M. K., Tan, K. T., Lee, K. T., & Mohamed, A. R. (2009). Malaysian palm oil: Surviving the food verses fuel dispute for a sustainable future. *Renewable and Sustainable Energy Reviews, 13*, 1456–1464.

Li, Y., Shi, Y., Zho, X., Cao, H., & Yu, Y. (2014). Coastal wetland loss and environmental change due to rapid urban expansion in Lianyungang, Jiangsu, China. *Regional Environmental Change, 14*, 1175–1188.

Lynch, M. J., Stetesky, P. B., & Long, M. A. (2016). A proposal for the political economy of green criminology: Capitalism and the case of the Alberta tar sands. *Canadian Journal of Criminology and Criminal Justice, 58*, 137–160.

Markham, K. E., & Sangermano, F. (2018). Evaluating wildlife vulnerability to mercury pollution from artisanal and small-scale mining in Madre de Dios, Peru. *Tropical Conservation Science, 11*. https://doi.org/10.1177/1940082918794320

Micklin, P. (2016). The future Aral Sea: Hope despair. *Environmental Earth Sciences, 75*. https://doi.org/10.1007/s12665-016-5614-5

Miller, D. (2012). *Last nightshift in Savar: The story of the spectrum sweater factory collapse*. McNidder & Grace.

Morelle, R. (2019, May 13). Mariana Trench deepest-ever sub dive finds plastic bag. *BBC News*. Retrieved June 1, 2019, from https://www.bbc.com/news/science-environment-48230157

Pierre-Louis, K. (2017, March 28). What the frack is in fracking fluid? *Popular Science*. Retrieved June 1, 2019, from https://www.popsci.com/what-is-in-fracking-fluid/#page-2

Reddy, C. M., Arey, J. S., Seewald, J. S., Sylva, S. P., Lemkau, K. L., Nelson, R. K., Carmichael, C. A., McIntyre, C., Fenwick, J., Ventura, G. T., Van Mooy, B. A. S., & Camilli, R. (2012). Composition and fate of gas and oil released to the water column during the Deepwater Horizon oil spill. *PNAS, 109*, 20229–20234.

Roberts, H. H. (1997). Dynamic changes of the Holocene Mississippi River Delta plain: The delta cycle. *Journal of Coastal Research, 13*, 605–627.
Schumacher, E. F. (1973). *Small is beautiful: A study of economics as if people mattered.* Blond & Briggs.
Sellers, C. C. (2012). *Crabgrass crucible: Suburban nature and the rise of environmentalism in twentieth-century America.* The University of North Carolina Press.
Watson, I., Shelley, J., Pokharel, S., & Daniele, U. (2019, April 27). China's recycling ban has sent America's plastic to Malaysia. Now they don't want it—So what next? *CNN.* Retrieved June 1, 2019, https://www.cnn.com/2019/04/26/asia/malaysia-plastic-recycle-intl/index.html
World Bank. (2019). Fertilizer consumption. Retrieved June 1, 2019, from https://data.worldbank.org/indicator/ag.con.fert.zs
World Bank. (2020). *What a waste 2.0: A global snapshot of waste management to 2050.* Retrieved December 5, 2020, from https://www.worldbank.org/en/news/infographic/2018/09/20/what-a-waste-20-a-global-snapshot-of-solid-waste-management-to-2050

4

Our Ethical Responsibility

Introduction

Over the last century or so, many individuals and organizations have worked hard to try to "save the planet." Greenpeace, for example, has a mission to "…expose environmental problems and promote solutions that are essential to a green and peaceful future." But why is Greenpeace interested in creating a green and peaceful future? Why do any of us care about the future of our planet? Also, why do some of us not care at all? What drives our ethical framework around sustainability and the environment?

This chapter will look at the development of our modern environmental ethics and make the case that humans have ethical responsibility to try to live more gently on the planet. As will be seen, our philosophical approach to the environment has changed over time and will continue to evolve. Also, this chapter will ask a basic question. Why are we interested in creating a more sustainable world?

Anthropocentrism

Many of us act like humans are the most important things on the planet. We live our lives as if we are all that matter and do not care about the environment. This anthropocentric viewpoint is rooted deep in the human psyche, in part, because of the importance of one phrase in the Old Testament.

> And God said, let us make man in our image, and after our likeness and let them have dominion over the fish of the sea, and over the fowl of the air, and over the cattle, and over all the earth, and over every creeping thing that creepeth upon the earth.

The keyword in this verse is dominion. According to the Old Testament, man is to have total rights of ownership of the earth and all living creatures. This idea permeates the actions of humans over the span of history.

This anthropocentric viewpoint gives cover for people who pollute, kill endangered species, cut down the remaining rainforests, and destroy coastal ecosystems as they develop hotels. They feel that they have the right to have dominion over nature and do what they want to do regardless of outcome. Indeed, the outcome is something that they do not have to worry about because it is all part of God's plan. There is no guilt for bad environmental actions because they have been given divine right to do what they want. They have dominion over nature.

Of course, over time, scientific inquiry and the enlightenment have shown us that human agency can have negative consequences for current and future generations. As a result of this, those who are fully committed to their right to have dominion over nature have rejected science and consider issues like climate change and environmental pollution unimportant or "fake news." Al Gore used the term, inconvenient truths, for the types of information that anti-enlightenment advocates disparage.

Anthropocentrism in many ways is in direct opposition of the environmental and sustainability movements. Sustainability is all about finding ways to preserve resources for future generations while anthropocentrism is all about humans first at all times. Anthropocentrism is closely related to ethnocentrism in that ethnocentrism also focuses on the importance of place and ethnic connection. Ethnocentrists may care about the environment, but only within the context of their local community or ethnic group. They reject global solutions and care only about their own community. They might be interested in solving environmental problems only if it benefits them.

Many believe that these anthropocentric and ethnocentric viewpoints are the reasons our planet has lurched into unsustainable conditions. We do not have enough long-term altruistic thinking to fully address our planetary environmental problems. However, as we will see, a range of ethical considerations have emerged in the last two decades that created a potent counter to traditional anthropocentric viewpoints. As we face a range of existential sustainability issues, anthropocentrism is seeming very antiquarian and more and more people are embracing an environmental ethic. It is important to note that Western traditions place man as separate from nature, while many Eastern traditions seek to place humans as closer in harmony with nature.

Preservation or Wise Use

One of the larger questions associated with our environmental ethic is how we interact with the natural world. Do we use it or do we preserve it? This question has been a fundamental question that has faced human society for generations. Of course, there is a continuum of options between use it or preserve it, which will be discussed shortly. However, it is worth considering what is meant when advocating preservation of the environment or the use of the environment (Fig. 4.1).

Preservation of the environment is the protection of a landscape and all that it contains without any impact of human activity. The goal is to allow nature to exist and thrive while avoiding the deleterious effects of man's agency. Nature preserves are often the expression of the preservation movement. In the United States, national parks are often held up as examples of the types of places where preservation ideals are in place.

While protection of nature is a laudable goal, the preserve movement does have some challenges. First, most of the natural world evolved over the last several thousand years in concert with humans. Thus, to remove

Fig. 4.1 This old railroad track has been turned into a walking trail and some aspects of a natural ecosystem have returned. How we use land over time has an impact on the environment

all human agency from a preserve does not truly reflect the character of a natural landscape. Clearly, the modern impact of humankind is not desirous either. However, the preserves of today do not fully reflect the natural world of the past.

Second, preserves often serve as refuges of nature within a highly altered landscape. While they serve to protect certain ecosystems, they are often not connected to other natural systems. They are islands of protection. As a result, these systems can end up with unique problems. For example, the Florida panther has developed genetic oddities such as a crooked tail due to the fact that their limited population is highly isolated within the Everglades National Park and other preserves in south and central Florida.

Preservation of landscapes and ecosystems is obviously better for the environment than their use. However, the use of any landscape can vary over space and time. Strip mining or mountaintop removal mining, which has taken place in coal-rich areas of the world, do tremendous damage to ecosystems because they are essentially removed. Other activities, such as farming, or housing development, are nearly as bad. However, there are other less damaging land uses such as forestry or golf courses. Thus, there is a continuum of human activity that disrupts natural systems.

What many have advocated is the wise use tenet that posits that natural land should be used for the benefit of humankind while trying to protect ecosystems and natural environments as much as possible. The challenging idea here is how to balance the benefits of humankind with the protection of ecosystems and natural environments. A mining company could argue, for example, that coal is a great benefit to humanity because it provides electricity. S/he might claim that the destruction of land to produce this benefit to humankind is worth the cost. Others would argue the exact opposite and that the value of the natural environment far outweighs the shorter-term benefit of utilizing coal in electrical generation. The wise use tenet, therefore, requires a strong framework and rule of law for decision-making.

The wise use and preservationist approaches came to a head in the early twentieth century when San Francisco sought to build a dam to bring water to the city after the great 1906 earthquake and fire (Righter, 2005). At the time, there was a great need to expand the city's water supply due to growing populations and because the city was unable to adequately fight the fires that started after gas lines ruptured during the earthquake. The city and state found a source of water in the Hetch Hetchy valley of the Tuolumne River located in the relatively new Yosemite National Park (established 1890) and sought to dam the river to secure water for their water supply.

Because the site was in a national park, building any kind of dam required an act of Congress. The US National Park system was established with the explicit purpose of protecting nature. Building a dam on national park property would change the mission. When the plan was announced, John Muir, as head of the newly established Sierra Club, sought to fight the dam. Muir, who worked hard to advocate for preservation, saw the dam as an encroachment of nature. However, others, including founding head of the US Forest Service, Gifford Pinchot, advocated for the building of the dam, arguing that the destruction of a small portion of the park was a small price to pay for the great benefits afforded to society by the construction of the dam.

Eventually, the US Congress and President Woodrow Wilson approved building the dam. It was the last time a major project like this was undertaken in a national park. However, the Hetch Hetchy controversy remains an example of a fundamental dichotomy of environmental ethics: to protect and preserve nature at all costs or to use nature for the greater good of humankind.

The Land Ethic, the Commons, and Environmental Law

Of course, most land does not fall under government protection and it is then up to owners or users of the land to decide how it is to be managed. Over time, individual owners and users of land have had mixed success with successfully managing land. Garrett Hardin points to the challenges of unplanned land use in his essay, The Tragedy of the Commons (Hardin, 1968). This piece highlights how shared land resources, such as a pasture, can be damaged if one person or groups of people overuse land in haphazard ways. It demonstrates that land must be carefully managed in order for it to be productive.

The idea of land management has been around for centuries as individuals and communities sought to tend their lands for future generations. However, as industrialized agriculture and profit-motivated activities promoted short-term gain over long-term land management, it became evident that we needed to have a much more nuanced view of how best to tend to our planet. Patterns of bad land management emerged in the twentieth century throughout our planet. The famous Dust Bowl in the US Great Plains in the 1920s was but one example of poor land management. However, widespread soil erosion throughout the world has created a range of agricultural challenges.

Soil erosion is not the only problem associated with bad land management. Loss of endangered species, eutrophication of waterways, and groundwater pollution are among many other symptoms of bad environmental decision-making. Entire ecosystems have been lost in this era. The Aral Sea disaster, for example, caused the loss of one of the world's largest bodies of water and led to widespread pollution in Kazakhstan and Uzbekistan. Likewise, the world has lost about 50% of its wetlands over the last century. These unique features are among the most productive ecosystems. They protect coastlines and help to clean surface water. In some areas, such as the Mississippi River Delta region, wetlands are being destroyed due to global climate change.

Over the years, we have come together to try to address bad land management decisions by creating rules and laws that regulate land use and activities (Brinkmann, 2020). We have done this at different scales of government. Internationally, the United Nations and other organizations developed binding agreements that require nations to limit activities that are harmful to the environment. There are many examples of these types of agreements; however, the Convention on International Trade in Endangered Species of Wild Fauna and Flora (CITES) is one of the most successful and well-known agreements. The international agreement, forged by the International Union for Conservation of Nature, went into effect in 1975 to limit the trade in endangered species. The agreement requires nations to develop policies to enforce trade in endangered species and also to develop penalties for illegal importation and use of them. The goal of the agreement is to protect endangered species in the wild. CITES is seen as a highly successful project since most of the nations in the world are signatories of the agreement.

National and local governments also developed laws to protect the environment in recent years. In the United States, a series of far-reaching laws were passed by the US Congress in the 1960s and 1970s including the Clean Water Act, the Clean Air Act, and the Endangered Species Act. Each of these laws focused on trying to protect the national commons—our air, water, and ecosystems. Likewise, local governments created their own laws that ranged from land use regulations to waste management and recycling requirements.

In many ways, these laws emerged from the writings of Aldo Leopold and Rachel Carson who highlighted that the environment was facing serious ecosystems decline in the industrialized twentieth century. Rachel Carson noted the problems associated with the unregulated use of industrial chemicals (Carson, 1962) and Aldo Leopold (Leopold, 1949) highlighted poor decision-making of public and agricultural lands. Taken together, their work is an argument advancing the need for the laws that started to emerge in the second half of the twentieth century.

Laws are an expression of a culture's collective ethics. We have demonstrated through the development of laws that we have adopted something that Leopold called a *land ethic*. We have come to recognize that we have a need to have an ethical relationship with nature. We cannot have a biblical approach that asks us to have dominance over nature. We have come to understand that unchecked dominance comes at a profound cost to the environment and to ourselves due to our many linkages with the environment.

Ecosystems and Modern Pollution

Our growing technological advancement has created new environmental problems that were not anticipated by those who developed the first wave of rulemaking around the land ethic. In the 1960s, we were concerned with a range of air and water pollution problems such as lead, ozone, and nitrogen. However, we have created new classes of pollutants and we have not fully come to terms with them. They are worth considering from an ethical standpoint as our societies contemplate how to manage them.

Pharmaceuticals. The use of pharmaceuticals has grown in recent decades. When we take them, traces of them are passed through our body and can enter ecosystems via sewage runoff. Most modern sewage treatment plants are unable to remove pharmaceuticals from the waste stream. In addition, many farm animals are injected with pharmaceuticals like hormones or are fed them within their diet. These chemicals can also pass through animal waste and enter ecosystems through agricultural runoff. There are many examples of pharmaceutical impacts on ecosystems.

Hormone pollution has impacted amphibian populations and trace levels of anti-depressants are making their way into water supply systems.

Plastics. Plastic pollution has emerged as one of the most vexing environmental problems of our modern era. Because many plastics take centuries to fully break down in ecosystems, they have built up in many areas to cause great harm. Most sea turtles and coastal birds have ingested plastics and there are many documented cases of marine mammal deaths caused by plastic. Small plastic particles have also entered our drinking water and food systems.

Personal Care Products. There are many types of personal care products that can cause environmental problems. They include things like lotions, makeup, and soaps. Because their use and diversity is increasing, there have been increased levels of these materials that have entered ecosystems through wastewater. A good example of an emerging problem is sunscreen. Thousands of tons of sunscreen are used each year by beachgoers. The sunscreen can wash off when swimming and the chemicals released into the water. Many areas have banned the use of certain types of sunscreen due to the environmental damage that they can cause.

Sound and Light. We have long known that sound and light pollution can disrupt ecosystems. However, we have not developed clear regulations around them in most areas of the world. Many marine animals rely on sound to communicate and all ecosystems evolved around the natural light of the sun, the moon, and the stars. There is emerging concern over the use of advanced sonar in military applications and their impact on marine mammals. Likewise, our world has brightened over the last 150 years and we have seen impacts on a range of plants and animals.

These are but a few of the modern land use and pollution issues that we have not fully addressed as a society. How we manage them into the future may very well decide our environmental future.

Problems with environmental pollution have revealed fundamental issues with environmental racism. Not all people on the planet are subject to the same levels of hazardous materials. There are many examples, most notably Flint, Michigan, that demonstrate that minorities are exposed to disproportionately high levels of pollution and environmental contamination. These issues, highlighted by Robert Bullard in the 1990s, demonstrate that even though environmental laws exist, they are not applied equitably in all areas. There are ethical failings within our society that often relate to race, ethnicity, and other differences.

Deep Ecology

Many question our ability to effectively manage environmental problems utilizing the ethical frameworks and laws we developed over the last century. They note that some key environmental issues like climate change or eutrophication of coastal systems have become existential for human society. Even though we have developed international environmental agreements like the Paris Climate Accord, the world continues on its unsustainable path. Critics of the ethical and legal approach to managing our environmental problems advocate for a return to greater connection to our planet. The thinking around this has been framed within the work of deep ecologists (Sessions, 1995).

The deep ecology school of thought emerged in the 1970s, in part as a reaction to the development of sweeping environmental regulations. While deep ecologists for the most part applauded the codification of an environmental ethic, they also recognized that the laws were flawed and could be corrupted. This idea was highlighted in the writing of Edward Abbey's groundbreaking books, *Desert Solitaire* and *The Monkey Wrench Gang*. Both books noted that government management of public lands could lead to their destruction. During this time, a range of industries challenged many of the environmental rules that came about during the 1960s, thereby weakening them. For example, even though many ecosystems were protected by these laws, many saw significant decline, most notably the loss of wetlands in the last half of the twentieth century.

Deep ecology suggests that the imperfect nature of environmental laws requires a deeper look into the relationship of humankind with the environment (Fig. 4.2). Looking around, deep ecologists saw that the imprint of humans was everywhere and that we needed to radically change our ways. As a result, many deep ecologists advocated for reducing our population through population control and for radically transforming our economic and social systems to ones that promoted the healing of our planet rather than its destruction. Environmental education is also a

Fig. 4.2 This prairie preserve in Illinois has been intentionally restored to bring back the native prairie ecosystem. The restoration shows that people care about the environment and our interaction with it

major theme of this philosophy. Humans have lost their connections to the natural world and we all need to know much more about it. Deep ecology, at its core, takes a scientific and spiritual approach to nature that highlights knowledge about how nature works and how it is altered by human activity, while also recognizing the deep spiritual connections that can be had by experiencing nature.

Activists like Julia Butterfly Hill emerged from the thinking and the activism of deep ecologists. Hill, who sat on top of a redwood tree for two years to protest the deforestation of old growth forests, had a deep connection to the environment and sought to protect it from legal timber production. These types of protests were highlighted in Abbey's book, *The Monkey Wrench Gang* (Abbey, 1975), and are examples of the way that environmental activists found new ways to protest the destruction of the environment with government complicity.

In many ways, some of the new environmental movements, like Extinction Rebellion, have their roots in deep ecology. Extinction Rebellion, for example, advocates civil disobedience around issues of climate change and biodiversity loss. Led largely by academics, the movement has held events that disrupt day-to-day activities in major global cities. The School Strike for Climate youth movement can also be traced to ideas that emerged around the deep ecology movement. Greta Thunberg, one of the youths involved with popularizing the movement, has advocated for new approaches to handling climate change since world leaders have been unable to enact meaningful change.

Ecofeminism

Another important environmental ethical movement is ecofeminism, which emerged around the same time as deep ecology (Warren, 1997). Ecofeminists assert that the current materialistic state of affairs is contrary to natural ecology. We live in a time with great resource extraction to support our modern consumerist society. The competition to amass greater wealth and power are masculine traits and these traits are the very ones that have led to the earth being out of balance.

Ecofeminists, like deep ecologists, call for resistance to the status quo and a return to greater connections to natural systems. Noted author Vandana Shiva has argued that women have much stronger connections to the earth and that women must reassert their roles as carriers of knowledge of the ways of nature. The ecofeminist approach closely examines power roles in the creation of our current problems. Ecofeminists note the significance of dominance themes in traditional religions as detrimental to the environment. Dominion (over nature), a male trait, is found in many foundational religious texts that have influenced generations. As a result, these male traits have taken precedence since their emergence to create a highly competitive and disruptive society. Competition for resources and dominance not only over the earth but over each other creates constant conflict.

Ecofeminists argue that we need to return to a time when there is more concern over the long-term care of the planet. This cannot be accomplished within our current capitalist system that rewards extraction and dominion. Instead, new economic approaches need to be developed. Some have argued that ecosocialism, which focuses on enhancing collective ownership, reducing capitalism, and restoring ecosystems, is the feminist path toward repairing the ills of capitalism (Albritton, 2019).

Another important aspect of ecofeminism is a spiritual aspect that suggests that modern religions have moved us out of balance with our home planet. Plus, as James Lovelock and others have pointed out, the earth can be seen as a self-regulating organism (Lovelock & Margulis, 1974). Some spiritual ecofeminists argue that we need to return to the ancient forms of religion, such as paganism and Wicca, which build greater bonds with people and the environment. By rebuilding these connections, we will have greater compassion for the planet and the environment and thus do less damage to it over the span of our lives.

Animal Rights

In recent decades, many have advocated strongly for enhancing the rights of animals (Garner, 1996). The public's interest in this topic derives from a variety of well-publicized issues associated with animals: species loss, hunting, zoos, the exotic animal trade, meat and fur production, animal testing, and the use of animals for pets.

Some of the most striking images of the last half a century have been of the last of a particular kind of animal that is going extinct due to habitat loss. We have all seen the images of the last few pitiful animals in zoos or refuges that tug the heart. However, these are but a few of the challenges we face. We are in a broader era of extinction throughout our planet as a result of human action. The loss is particularly accented during our era due to unethical hunting practices. Most hunting around the world is done ethically within the aegis of local, state, and national regulations. However, there are unethical groups who promote trophy hunting of endangered and threatened species which has caused a tremendous worldwide outcry in support of these rare animals.

Zoos, both public and private, have also come under scrutiny. One of the most popular documentary series of 2020 around the world, *Tiger*

King, focused on private big cat zoos. Big cats, most notably tigers, are bred for not only zoos, but the exotic animal trade for private collectors. While many public zoos focus on protection and preservation of wild animals, the documentary highlighted how private zoos are linked to the unethical trade of exotic and endangered species. Some of these exotic animals are also linked to a range of traditional medicinal applications which further complicates the issue.

Of course, one of the largest issues associated with animal rights is meat and fur production. Globally, there is a hunger for meat. People around the world are eating more meat as they have become more affluent. This carnivorous drive is supported by the growth of industrial-scale agriculture that produces large amounts of meat efficiently—although many have criticized this efficiency as an unethical way to produce meat. There are many examples of the types of poor living conditions that exist for animals stuck in the industrial meat system. Most of the criticism derives from the cramped living conditions for the animals. However, many have noted that workers lose their humanity when having to manage the growth and eventual butcher of these animals.

Fur production has also faced considerable criticism. A widely successful campaign around fur managed by People for the Ethical Treatment of Animals (PETA) has made the wearing of fur a political statement. In the last few decades, widespread use of fur has fallen out of fashion. It is still used by some luxury designers, but you rarely see a fur store as one might have seen in the past. Nevertheless, in many parts of the world, thousands of animals are still raised for the fur trade and there is still some wild trapping that takes place to produce wild fur for the market.

Another area that has received great attention is laboratory testing using animal subjects. Here there are a range of ethical issues. Certainly, testing life-saving medicine is different than testing personal care products like shampoo or skin cream. However, to some, any animal testing is unethical and should be banned. There is no doubt that the lives of most laboratory animals are grim. Over the years, in part due to pressure from animal rights activists, organizations have implemented rules for the managing of laboratory animal facilities. For example, Institutional Animal Care and Use Committees are typically in place in most institutions that utilize animals for laboratory testing. These committees evaluate the need for each animal study and also work to ensure that the animal facilities are maintained appropriately. To ensure animal safety, the committees have at least one veterinarian member, and to ensure that the institution is meeting community standards the committees have at least one member from outside of the institution.

One other animal rights theme that has emerged in recent years is the whole notion of animals as pets. Archaeological evidence has shown that people have lived with animals for millennia. We have long had friendly relations with a range of animals, most commonly cats and dogs. In recent decades societies around the world have increased the number of pets they live with in their homes. The pet industry is big business. Of course, this has ushered in a range of critiques about pet "ownership" that leads to interesting ethical questions. First of all, can we truly "own" pets? Many have opted to utilize the term "companion animal" to designate the animals with which they live. The growth in the pet industry has led to highly selective breeding programs that are not always that healthy for the animals. Plus, there are many instances of unethical puppy mills involved with raising animals under unhealthy conditions. Plus, we all have heard of the people who end up taking in too many animals who are unable to care for them or people who defend letting non-native house cats run free even though they are responsible for decimating the songbird population.

As mentioned earlier, one of the leading organizations that spearhead many animal rights educational initiatives is PETA. Started in the United States in the 1980s, PETA is known internationally for advocating on issues ranging from veganism to cock fighting. They have brought forward many ethical issues to the general public via thought-provoking advertisement campaigns utilizing well-known celebrities like Pamela Sue Anderson.

We all have some type of relationship with animals. Choosing to see animals as food, companion animals, pets, exotica, lab creatures, clothing, or as unintended casualties of modern life helps define us as human beings. We all have an animal ethic—but some of us make vastly different ethical choices around animals that can lead to societal tensions.

Applied Theology

Another important aspect of environmental ethics is applied theology which seeks to apply religious teaching to the reality of human existence. Many of the main religions of the world have embraced environmentalism. For example, the Pope's Climate Covenant asked Catholics around the world to better understand the role of faith in environmental protection (Wilkins, 2020). According to the Pope, the destruction of the environment due to the actions of humanity is fundamentally an ethical issue that needs to be addressed by faith.

The Catholic Church is not the only faith-based organization to address modern environmental issues. Most Christian, Muslim, Jewish, Buddhist, and other religious organizations have made the environment a major theme of their teachings in recent years. Many organizations within congregations have formed to teach and advocate on environmental issues—particularly climate change. For example, Young Evangelicals for Climate Action work within evangelical churches and college campuses across the United States to try to encourage action on climate change. They highlight that protection of the environment is a fundamental Christian tenet and make it a cornerstone of their faith.

Across the world, as governments are failing in their duty to protect the environment, religious organizations are taking leadership roles. They, along with other leaders in the business and non-profit world are influencing their stakeholders to live a more ethical life in harmony with nature. They are utilizing their influence to create new ethical understanding around nature and the environment.

Business Ethics

As religious leaders have moved their congregations to understand that protection of environment is part of religious ethical teaching, business leaders have also embraced an environmental and sustainability ethic. Business ethics in the past focused around issues like honesty, fraud reduction, and fairness. However, in recent years, business leaders have embraced a range of ethical issues including human rights, environmental protection, and sustainability.

The range of commitment to sustainability within the business community varies considerably. Some companies produce annual sustainability reports that review all of their operations within a sustainability lens. They provide baseline data, set goals, and report on goals within the three E's of sustainability. Other companies may pick one aspect of sustainability, say greenhouse gas reduction, and focus their attention on that area.

Because of the high variability in sustainability reporting, a number of organizations provide external guidance, support, and verification of corporate sustainability efforts. The Global Reporting Initiative (GRI), for example, provides a range of guidance on social, environmental, and economic indicators (GRI, 2020). Formed in 1999 and influenced heavily by the work of the United Nations, the Initiative has worked with the majority of the top international organizations to guide them into advancing ethical sustainability goals. Likewise, the International Organization for Standardization (ISO) developed sustainability and human rights standards that many organizations are using (ISO, 2020).

Certainly, these two forms of reporting, GRI and ISO, do not fully address all ethical aspects of business activities. Some of the organizations that utilize these reporting tools have engaged in unethical business practices in recent years. However, they do provide ethical guidance within the confines of sustainability as defined within the three E's of environment, equity, and economy.

Why Care?

At the close of this chapter, it is worth considering why any person, business, or religious organization should care about environmental ethics and sustainability. Indeed, protection of the environment sometimes comes at personal costs or loss of business. Some may argue that, at least in the realm of business, that such caring is artificial—it may be considered a new form of marketing or greenwashing activities. There may be something to this. However, I would argue that caring about the environment within an ethical framework is a return to basic human ethics. Perhaps that is why you are reading this book?

References

Abbey, E. (1975). *The monkey wrench gang*. Lippincott Williams & Wilkins.

Albritton, R. (2019). *Eco-socialism for now and the future: Practical utopias and rational action* (Palgrave Insights into Apocalypse Economics). Palgrave Macmillan.

Brinkmann, R. (2020). *Environmental sustainability in a time of change*. Palgrave Macmillan.

Carson, R. (1962). *Silent spring*. Houghton Mifflin.

Garner, R. (1996). *Animal rights: The changing debate*. Palgrave Macmillan.

GRI. (2020). Welcome to GRI. Retrieved December 5, 2020, from https://www.globalreporting.org

Hardin, G. (1968). The tragedy of the commons. *Science, 162*, 1243–1248.

ISO. (2020). ISO 26000—Social responsibility. Retrieved December 5, 2020, from https://www.iso.org/iso-26000-social-responsibility.html

Leopold, A. (1949). *A Sand County almanac*. Ballantine.

Lovelock, J. E., & Margulis, L. (1974). Atmospheric homeostasis by and for the biosphere: The gaia hypothesis. *Tellus, 26*, 2–10. https://doi.org/10.3402/tellusa.v26i1-2.9731

Righter, R. W. (2005). *The battle over Hetch Hetchy: America's most controversial dam and the birth of modern environmentalism*. Oxford University Press.

Sessions, G. (Ed.). (1995). *Deep ecology for the twenty-first century: Readings on the philosophy and practice of the new environmentalism*. Shambhala Publications, Inc.

Warren, K. J. (1997). *Ecofeminism: Women, culture, nature*. Indiana University Press.

Wilkins, D. (2020). Catholic clerical responses to climate change and Pope Francis's Laudato Si'. *Environment and Planning E: Nature and Space*. https://doi.org/10.1177/2514848620974029

Part II

Tackling Climate Change

In November 2020, Hurricane Iota hit coastal Nicaragua in nearly exactly the same spot as Hurricane Eta did a few weeks earlier. The back-to-back storms killed hundreds of people and devastated the mountainous country with landslides and flooding. The storms highlighted that we live in a world that is profoundly changed. Our hurricanes are getting more intense and deadly due to climate change. All of our natural systems are in some way altered through our actions. This section of the book highlights how we can take direct action to address climate change. Chapter 5 focuses on how we as individuals can reduce our impact and calculate our carbon footprint. Chapter 6 looks at issues of community and governmental approaches to greenhouse gas reduction. Climate change policies associated with businesses, schools, and non-profits are reviewed in Chap. 7. Each of these chapters provides rich reviews of what we and our organizations can do to reduce greenhouse gases and slow the impacts of climate change.

5

How You and Your Family Can Reduce Your Carbon Footprint

Introduction

I think by this part of the book we can all agree that reducing your carbon footprint is good for the planet. In this chapter, we will explore the ways that we can calculate the amount of greenhouse gases we produce annually. As will be seen, it is really not that hard to figure out the impact we have on our planet. The chapter will also review the sources of our greenhouse gases and from where they come. The chapter will specifically focus on greenhouse gases produced by heating and cooling, transportation, food, and waste and how we can reduce their production in each of these categories. Finally, the chapter will highlight how we can mitigate our carbon footprint by purchasing carbon credits or creating carbon sinks on our own.

It would be easy to not concern oneself with one's carbon footprint given the magnitude of the problem. However, one person truly can make a difference. All over the planet, over many years, individuals made decisions that created our current greenhouse gas and climate change problems. Today, many individuals are changing their ways to make wiser decisions that benefit our planet. You can choose to live in the now and be part of this collective solution to the problem or live in the past and continue to be a burden on the environment. Each of us working individually to reduce our greenhouse gas emissions helps to make a meaningful change.

As you work your way through this chapter and the ones that follow that have a similar format, it is worth considering how far you want to go to create positive change. Throughout the remainder of the book, there are tables that provide a range of options for you to consider as you transform your life to live more sustainably. Some of the options are simple and some of them are quite hard. It may be worth considering at the outset how much you want to transform your life. Are you interested in having the biggest impact with the smallest real lifestyle change? Or are you interested in a fundamental lifestyle change that includes a vastly more sustainability impact? Most of you probably fall somewhere in between. No matter where you are in the continuum, it is worth considering how you can challenge yourself to make more difficult choices to live more sustainably.

Carbon Footprint Calculations and Carbon Dioxide Equivalents

Chapter 2 reviewed the range of greenhouse gases that are impacting our environment. At the personal level, the two most common greenhouse gases that we produce are carbon dioxide and methane. Carbon dioxide is produced in the burning of fossil fuels and methane is produced through the decomposition of organic matter and through leaks of natural gas lines, including those that make their way into our home. Given that we are mainly concerned with just two basic chemicals, it is relatively easy to calculate how much of them we produce through our activities over a span of time. Most carbon calculators focus on annual production of greenhouse gases, but it is important to recognize that there are temporal variations that emerge due to things like climate and purchasing or vacation trends. For example, for many of us, our greenhouse gas footprint increases significantly over holidays due to increased purchasing.

5 How You and Your Family Can Reduce Your Carbon Footprint

Because of the nature of how we individually use greenhouse gases, we can break down the calculation of our carbon footprint into three main categories: housing, transportation, and secondary sources such as those produced from food consumption and a variety of other sources. The first step in assessing our footprint is to pull together information based on our annual consumption. Following this, we can enter our data into one of several online carbon footprint calculation tools.

Data Collection. In order to calculate the impacts of our housing, we need to calculate how much energy we consume from different sources. In my home, for example, I use electricity for heating and cooling and natural gas for cooking and for clothes drying. Around the world, there are a variety of sources of energy used to produce electricity and thus different regions have different greenhouse gas impacts. Two different homes in different regions with the same usages of kilowatt-hours of energy can have vastly different impacts based on how the energy is produced. One home, for example, may be on an electrical grid where the majority of the electricity comes from coal, while another may be on an electrical grid where the majority of the electricity is produced from renewable sources. As a result, most utilities now provide a greenhouse gas conversion factor, often given in the unit of $kgCO_2e/kWh$, that must be used to assess the overall climate impact of electrical use.

Thus, in order to calculate the impact of electrical use in the home, you need to find your annual kilowatt-hours of energy used over the span of a year and you need to know your greenhouse gas conversion factor. If you cannot find the conversion factor, many greenhouse gas calculators that you can find online (more about that soon) utilize the national average of the conversion factor which is probably close enough for your purposes.

Most homes utilize other energy sources besides electricity (Fig. 5.1). As I noted earlier, I use natural gas which is reported in my energy bill as a certain number of therms of usage every month. Many homes use heating oil which can be measured in gallons or some other volume and others use propane, measured in volume such as gallons, or wooden pellets, measured in weight such as tons. Regardless of the energy sources used in your home, in order to calculate your greenhouse gas footprint, you need

Fig. 5.1 Calculating your greenhouse gas footprint can be complicated. For example, the people living in these condominiums and apartments along this Vancouver waterfront have to consider things like heating, cooling, lighting, transportation, waste, and food sources

to calculate the annual amount of energy you use to power, heat, and cool your home.

Transportation, the second of the three categories for greenhouse gas footprint calculations, is often rather complicated because of the variety of transportation choices we have for travel and for the temporal variability of travel throughout the year. If you are a car owner, it is relatively easy to calculate the impacts from your driving. If you have purchase records, you can total the number of gallons you bought to fill up your car. However, most of us do not keep such detailed records and you can easily assess the number of miles you drove over the span of a year from your odometer and utilize your miles per gallon to calculate the number of gallons of gasoline you used. For example, if you drove 8000 miles over the span of a year in a car that gets 40 miles to the gallon, you used

5 How You and Your Family Can Reduce Your Carbon Footprint

approximately 200 gallons of fuel. Different fuel types have different greenhouse gas impacts per gallon burned and most greenhouse gas footprint calculators take this into consideration.

The impacts of other forms of ground transportation are more challenging to calculate due to the high variability of impacts and the lack of access to data on local greenhouse gas emissions from buses, trains, trams, subways, and taxis. As a result, most greenhouse gas footprint calculators take the national average of emissions from these different sources. However, you will need to calculate the number of miles traveled over the span of a year using these different sources. For example, many of my friends take the subway in New York City about 5 miles each way during their commute to work. As a result, over the span of a year, they travel roughly 2500 miles each year by subway (assuming a two-week vacation and a five-day workweek).

Air travel is especially challenging to calculate, but it is important to include it into footprint calculations because of its significant impact. For those of us who travel by air quite a bit, the calculation can be especially difficult. Most carbon calculators require one to enter each individual flight—which can be quite a bit of work if you travel for work. However, most of us will need to enter a flight or two (if any). Some calculators are more general and require you to just enter the number of flights you have taken and the approximate miles traveled. Regardless of the approach, the actual carbon calculations are rough approximations. The actual emissions depend very much on how many people were on the flight and the weight of the aircraft.

It must be stressed that it is important to include air travel in greenhouse gas calculations. One round-trip flight from New York to China often has the same emissions as someone's annual automobile commuting. We often underestimate the impact of our air travel when we think about our greenhouse gas impact because we utilize lower impact forms of transportation, such as cars and buses, more frequently.

Secondary sources are the third major category of greenhouse gases that we produce. This is a catch-all category and includes a range of sources such as food, waste, and the types of products we purchase. As you can imagine, your lifestyle and income contribute quite a bit to the quantity of greenhouse gases produced in this category. A vegetarian, for example, has a much lower greenhouse gas footprint than a meat eater. Someone who purchases lots of clothing and other consumer goods also has a larger greenhouse gas footprint than someone who lives more frugally.

Crunching the Numbers. Once you have all the data you need to calculate your greenhouse gas footprint, you need to crunch the numbers to arrive at a grand total. There are a number of published carbon footprint calculators that one can easily find online. Some are relatively simple (The Nature Conservancy, 2019) and ask you to enter a modest amount of data while others are more complete and thus more accurate. This is not an endorsement, but one that I like to use is one published by Carbon Footprint Ltd at carbonfootprint.com/calculator.aspx (Carbon Footprint, 2020). It is comprehensive in the types of information one can enter and it thus provides a relatively accurate assessment of your total carbon use. Unfortunately, it does not take into account carbon offsets which will be discussed later in the chapter.

Regardless of the carbon footprint calculator you opt to use, the key outcome is a total amount of carbon dioxide equivalent you produce each year. What is important to note is that this is the total amount that you are responsible for in your home. It does not take into account the carbon you may be responsible for as a result of your job. Many companies conduct greenhouse gas inventories to assess their carbon footprint. Once these inventories are complete, they often work to reduce their footprint or find ways to offset their greenhouse gas impacts through the purchase of carbon offsets. It is worth checking with your employer to find out their policies on greenhouse gases to assess if your carbon contribution on the job is addressed in some way.

The other thing to remember when calculating your carbon footprint is that some of the data you may be entering represent the total energy usage for your home. In order to get your personal greenhouse gas footprint, you need to consider the number of people in the home and make the appropriate calculation. This issue highlights an important issue that is significant in sustainability discussions: The denser use of any built environment such as a home, city, or office building, will lead to lower per person greenhouse gas footprints. This is why densely populated places like New York City have a lower per person greenhouse gas footprint than sprawling places like Tampa, Los Angeles, or Phoenix.

Greenhouse Gas Equivalents. As was noted in Chap. 2, greenhouse gases have different properties that cause them to react differently in the atmosphere. Because of this, they each have different warming potentials. This means that in order to calculate our carbon footprints we need to develop units of measurement that allow us to total up the overall warming potential of all of our greenhouse gas emissions. We do this by using units called carbon dioxide equivalents. Thus, we compare all greenhouse gas warming potentials to the warming potential of carbon dioxide. Because methane is 25 times more powerful of a greenhouse gas than carbon dioxide, one molecule of methane is the same as 25 carbon dioxide equivalents.

Using carbon dioxide equivalents allows us to create a single number that represents our greenhouse gas emissions from all sources. Once this number is attained, reduction goals can be set. Of course, the best goal is to try to be carbon neutral. However, it would certainly be easy enough to reduce greenhouse gases by 10% or 25%. The following sections focus on how we can achieve our carbon reduction goals.

Reducing Our Carbon Footprint

Once we know our carbon footprint, it is relatively easy to set reduction goals and targets and follow up that goal setting with specific reduction strategies. The following section will focus on how to reduce our carbon footprint in the following areas: (1) heating, cooling, and electrical use, (2) transportation, and (3) food and consumer goods. As will be shown, it is relatively easy to make substantial reductions of 10% or 25%. In fact, as will be seen in the upcoming section on carbon sinks and carbon credits, it is relatively easy to become carbon neutral.

The average US carbon footprint is approximately 16 tons of carbon dioxide equivalents per person. Globally, the rate is much less and averages at around 4 tons per person. Clearly a 25% reduction in greenhouse gas use by every American is possible given that it is one of the highest per capita rates in the world. In fact, even a reduction of 25% to 12 tons of carbon dioxide equivalents per person is higher than the average per person rate in the European Union which is approximately 7 tons of carbon dioxide equivalents per person. Both of these places (the United States and the European Union) are far higher than the average per person rate in India which is approximately 1.5 carbon dioxide equivalents.

No matter where you live and no matter what your carbon footprint is—from the carbon heavy United States to the carbon lite subcontinent of India—we can all try to reduce our personal carbon footprint on the planet.

Reducing Carbon from Heating, Cooling, and Electrical Uses. Most of us heat our homes using electricity or natural gas. However, some of us augment our heating through the use of woodburning stoves. Regardless of the way you heat or cool your home, you should have been able to calculate your overall energy use for your home in carbon dioxide equivalents using the methods described in the previous section. Outlined in Table 5.1 are several ways that we can individually reduce energy use in the home broken down into the following categories: heating and cooling, hot water, and general electrical use. As will be the pattern throughout the book, the reduction options are broken down into simple, hard/expensive, and innovative/life-changing approaches.

The simple approaches to heating and cooling are ones that many know about, but few consciously apply in their daily lives. The approaches include adjusting thermostats, caulking windows, adding insulation film to windows, and a variety of other home weatherizing approaches. It is remarkable how much energy can be saved by just making a few small changes. Harder and more expensive options include replacing HVAC systems with a more energy-efficient one, installing a smart thermostat that regulates temperature based on when the home is in use, and replacing old drafty windows with new energy-efficient systems. The innovative and life-changing options are more complex and include redesigning the home to take advantage of passive solar energy, moving to a smaller home, and the installation of innovative heating and cooling systems such as geothermal or photovoltaic energy systems.

The range of options here, and throughout the book, gives us an opportunity to contemplate our commitment to sustainability, and in this particular case, climate change. The simple options give us an

Table 5.1 Simple, hard/expensive, and innovative/life-changing approaches to energy reduction in the home

	Heating and cooling	Hot water	General electrical use
Simple	Adjust thermostats Caulk windows Add insulation film to windows Weatherize the house	Reset temperature of water heater Do not use hot water for clothes washing Take shorter and cooler showers	Turn off computers and other equipment when not in use Change lightbulbs to LED Hang dry your clothing
Hard/expensive	Replace HVAC with energy-efficient system Install smart thermostat Replace windows with energy-efficient windows	Replace water heater with energy-efficient system Reduce the size of the hot-water heater Put in a smart system that programs the hot water to be ready only when you need it	Replace appliances with energy-efficient ones Use smart power strips to reduce electrical use
Innovative/life-changing	Utilize passive solar Move to a smaller home Install geothermal, photovoltaic, or other innovative heating/cooling system	Get rid of your hot-water system Install solar water-heating system	Reduce the number of electrical appliances you use Install solar or wind-energy electrical system

opportunity to make a modest difference. The simple options will result in anywhere from 10% to 25% reductions in energy use. This is great and anyone should be applauded for making these changes. However, real transformative change takes place when one embraces the harder, innovative, or life-changing options. People all over the world are committing to some of the more difficult choices seen in Table 5.1. Are you ready?

Hot water systems are responsible for approximately 20% of the electrical usage in the home. The simple options for water heating include resetting the temperature of your water heater, cutting the use of hot water for clothes washing, and reducing the time and temperature for showers and baths. The harder and more expensive options include replacing the water heater with more energy-efficient systems, reducing the size of the hot water heater, and using a smart system to program the heater to produce hot water only when needed. The most innovative or life-changing options are to get rid of your hot water system or to install a solar hot water heating system. The cheapest option, of course, is to do without a hot water system. While it may seem challenging to live without hot water, it must be understood that many people around the world do not have them and they live happy and productive lives.

We use electricity in a range of household appliances such as computers, dishwashers, exercise machines, lighting, and kitchen equipment. The simple ways we can reduce electrical use is to turn off and unplug computers and other appliances when not in use, hang dry your clothing, and replace lightbulbs with energy-efficient options. Harder and more expensive choices involve replacing older less energy-efficient appliances with more energy-efficient ones. In addition, utilizing smart power strips cuts off electrical use when electrical equipment is off or idle. The harder options focus on reducing or eliminating certain appliances in the home and the installation of wind or solar electrical systems.

Reducing Carbon from Transportation. The percentage of national greenhouse gas emissions from transportation varies considerably from country to country. In the United States, for example, transportation sources account for almost 30% of its greenhouse gas emissions, while in poorer countries with limited transportation infrastructure, like Yemen, the impacts from transportation can be negligible. Of course, driving some of the challenges with transportation is a global car culture which promotes personal transportation convenience over mass transit. Car ownership is increasing around the world. Some wealthy countries like the United States and New Zealand have nearly one car on the road for every person. Poorer nations like Sudan, the Philippines, or Honduras have only a few cars per hundred people.

There are a number of ways to reduce our greenhouse gas impacts as a result of our transportation choices. As can be seen in Table 5.2, there are simple, hard/expensive, and innovative/life-changing approaches that can be employed. If you utilize the simple approaches you will probably

5 How You and Your Family Can Reduce Your Carbon Footprint

Table 5.2 Simple, hard/expensive, and innovative/life-changing approaches to reducing greenhouse gases from our transportation choices

	Transportation
Simple	Replace traditional gas-powered car with hybrid vehicle
	Drive or take the train for domestic trips instead of flying
	Carpool and use mass transit when possible
Hard/expensive	Replace traditional gas-powered or hybrid car with electric vehicles
	To reduce your impact on national or international shipping, commit to purchasing locally when products are available
	If you live near your job, get rid of your car and commit to mass transit, walking, or biking
Innovative/ life-changing	Commit to cutting air travel from your life
	Relocate near your job to reduce the need for cars or mass transit
	Eliminate all online purchasing and only buy local

cut a modest 10–25% of your greenhouse gases from the transportation component of your carbon budget. However, if you want to be more aggressive and innovative, you can more or less cut your greenhouse gas emissions from transportation from your overall budget.

The easiest approaches depend somewhat upon where you live and the transportation infrastructure available to you. If you drive a car, the simplest way to reduce greenhouse gases is to drive a hybrid or join a carpooling team for getting to and from work. Such efforts could cut your gasoline usage by 25% or more. In addition, if available, you can leave the car behind and utilize mass transit. In addition, you can cut those energy-guzzling flights and drive or take a train to your next vacation destination.

More substantial and harder ways to cut your greenhouse gas emissions require a bit of a change of lifestyle. First, if you want to keep your car for some reason, you can replace your traditional or hybrid car with an electric car. If you live near transit lines you can get rid of your car altogether and take the bus or subway and walk or ride a bike to your destinations. You can also cut the impact of your purchases on the global transportation network by committing to buying only local products.

Some of the harder choices you can make can involve significant personal change. If you do not live near mass transit and want to live car free, you can relocate near your job or find a job that you can do from your home. Another thing that you can do is to commit to not purchasing online and reducing your consumption overall. Most of us live with an incredible amount of stuff that we purchase online utilizing a variety of e-commerce platforms. Our purchases have transportation emissions associated with them. You can also reduce all air travel from your life. Many people who are serious about reducing their greenhouse gas emissions have committed to cutting air travel. Perhaps the most famous example of this is the roundtrip voyage Greta Thunberg took by sea from Europe to New York to speak at the United Nations. The COVID-19 situation has shown us that we can move many meetings and events online. We do not need to travel to make international connections. We have the technology to replace travel with online experiences. There is no doubt that small and easy personal changes, such as using a hybrid vehicle have an impact, but substantive change happens when we commit to a major alteration of our lifestyle.

Reducing Greenhouse Gases from Secondary Sources (Food and Waste). There are a range of other secondary sources of anthropogenic greenhouse gases, but the two most common are those that arise from our food consumption and the waste that we produce (Haass et al., 2015). Agricultural greenhouse gases arise from soil chemical processes, animal flatulence, and the energy used in agricultural processes and transportation of food within a globalized food system. Significant attention has been given to animal flatulence because the methane released has 25 times the warming potential as carbon dioxide. Table 5.3 highlights ways that you can reduce your greenhouse gas impact from food and waste.

The three easiest ways to reduce your carbon impact from your food choices are to start a garden, go meatless once a week through meatless Mondays, and buy local produce at a farmers' market. If you want to commit in a deeper way, you can participate in community-sponsored agriculture (CSA) and join a farm cooperative. You can also develop your own protein sources by raising backyard chickens for eggs. There are a range of lifestyle changes you can embrace to significantly reduce your greenhouse gas impact such as transforming your garden to permaculture or expanding your backyard chicken production to help others eat locally. However, one of the most impactful things you can do is to embrace

5 How You and Your Family Can Reduce Your Carbon Footprint

Table 5.3 Simple, hard/expensive, and innovative/life-changing approaches to reducing greenhouse gases from our food choices and personal waste management

	Food	Waste
Simple	Meatless Monday Purchase at farmers' markets Grow produce in a garden	Reduce purchasing Buy in bulk Reduce food waste
Hard/expensive	Go vegetarian Join a CSA Backyard chickens	Compost Purchase only things that do not lead to waste
Innovative/life-changing	Go vegan Support vegan/vegetarian causes	Live a zero-waste life

veganism and promote it in your community. Many activist groups are embracing veganism as part of their mission. Food Not Bombs, for example, is a group that focuses on homelessness and hunger. They help feed the poor utilizing only vegan or vegetarian meals. We often make food choices for others in our families, when we host a dinner, or when we plan catered events. By making choices to focus on vegan options, we are not only reducing our greenhouse gases, but we are also helping to educate others about vegan food.

Waste is also a major contributor to greenhouse gases in that the decomposition of organic material leads to the production of methane. Methane can be a problem for some landfills in that it can lead to explosions or fires if not managed correctly. Indeed, some landfills mine their methane for energy use. We can all make a range of choices to cut our waste impact. Two of the simplest things we can do are to cut our purchases and to cut our food waste. Modern consumerist societies produce huge amounts of waste that are challenging to address. Plus, we buy too much food and much of it ends up in the waste stream. Some activists have advocated moving to a zero-waste lifestyle or fully transforming our lives so that our carbon emissions from our purchases are significantly reduced. The zero-waste lifestyle may seem difficult to achieve in our modern society, but it involves fully committing to rejecting consumerism and focusing exclusively on what we truly need to live a happy and fulfilling life.

Carbon Sinks and Carbon Credits

Carbon sinks are those things that actually store carbon. They help us mitigate greenhouse gas emissions and thus subtract from our overall total greenhouse gas footprint. There are many types of carbon sinks. The Amazon rainforest, for example, has been in the news as of late because of the extensive deforestation currently underway. Because of the devastation, we are not only losing a vast sea of carbon storage, but we are also releasing that carbon into the atmosphere as the vegetation and soil burn. Some are countering this through aggressive tree planting campaigns. Many countries in Africa, for example, are in the midst of transformative reforestation efforts.

Trees, while an obvious form of carbon sink, are not the only tool we have to store carbon. Soil is also an excellent way to store carbon. Unfortunately, due to soil erosion and bad agricultural practices, we have destroyed and continue to destroy natural soil systems that store carbon. The same is true of the world's wetlands. These unique ecosystems, including the soggy tundra of the far north, store huge amounts of carbon. When they are destroyed, plant material which has built up for centuries can oxidize over a relatively short time period. Thus, restoring and protecting soil and wetlands is a sound way for us to store carbon. Oceans also absorb huge amounts of carbon dioxide.

But how can we as individuals use carbon sinks to mitigate our own greenhouse gas footprint? The answer is both difficult and simple. It is difficult because it is very hard to store or sequester the greenhouse gases for which we are responsible. A typical tree, for example, stores less than 100 pounds of carbon per year. This can vary, of course, by tree type or age, but this number serves as an example for understanding the scale of the carbon storage issue. The average American is responsible for around 32,000 pounds of carbon dioxide equivalents per year. This means, that every American would need to have 320 trees mitigating their greenhouse gas footprint. It is not a matter of just going out and planting a single tree.

As a result of the magnitude of the issue, a number of organizations have cropped up over the last several years to help individuals and organizations mitigate their carbon footprint. They take on big projects like soil restoration, reforestation, and methane collection from landfills in order to store or collect large amounts of carbon dioxide equivalents. They then sell greenhouse gas credits to us so that we do not have to figure out how to mitigate our own impact.

The costs for mitigating your own carbon footprint are relatively small. The cost for mitigating the annual emissions of an average American is around $300. Most of these companies also provide a range of other services. For example, some provide services for large companies to calculate their footprint and then develop mitigation strategies. They will also provide opportunities to purchase credits for travel, vacations, and special events such as conferences or weddings. The bottom line is that there are opportunities for us to purchase carbon credits to mitigate our carbon footprint that remains after we have worked to reduce it using strategies outlined in Tables 5.1, 5.2, and 5.3.

There are a number of companies that provide the opportunity to purchase greenhouse gas credits. The one that I use is called Terrapass, but this is in no way an advertisement or endorsement (Terrapass, 2019). There are plenty of other companies that do a fine job in this area such as Second Nature and even Tesla.

Remembering the Purpose

Living in a time of climate and environmental crisis provides plenty of motivation to work hard to try to reduce and mitigate our greenhouse gas footprint. This chapter provided methods for calculating your greenhouse gas impact as well as strategies for reducing your footprint from home energy use, transportation, food, and waste. It also provided ways to mitigate your remaining greenhouse gas emissions after you have done all you care to do to cut your climate footprint. As you work through how to address the range of issues associated with reducing your overall environmental impact associated with climate change, it is important to recognize that you are doing this because the earth is deeply out of balance. You are joining a growing number of people who are seeking to change global culture so that we can stop global climate change. You are also recognizing that you have a distinct power to impact future generations not only through your greenhouse gas reductions but also through your example.

One of the major tenets of sustainability is managing resources now so that future generations can thrive. The last few generations have willingly overconsumed natural resources and damaged the earth's climate even though they knew the impacts of their actions. The actions that you take as a result of this chapter, due to the time scales associated with climate change, will have far-reaching consequences that we may never fully see or understand. However, all of the science tells us that we must take these actions and we must take them quickly.

References

Carbon Footprint. (2020). Carbon calculator. Retrieved December 5, 2020, from https://www.carbonfootprint.com/calculator.aspx

Haass, R., Dittmer, P., Veigt, M., & Lütjen, M. (2015). Reducing food losses and carbon emission by using autonomous control—A simulation study of the intelligent container. *International Journal of Production Economics, 164*, 400–408.

Terrapass. (2019). Terrapass. Retrieved June 1, 2019, from https://www.terrapass.com/

The Nature Conservancy. (2019). Carbon calculator. Retrieved June 1, 2019, from https://www.nature.org/en-us/get-involved/how-to-help/consider-your-impact/carbon-calculator/?gclid=EAIaIQobChMIiY_MiqWs4QIVko7ICh1NnweYEAAYASAAEgKKb_D_BwE

6

How Your Community Can Reduce Its Greenhouse Gas Impact

Introduction

Human settlements of all sizes are starting to confront the realities of the climate challenge we face. For some coastal communities, like Miami, Florida, or Venice, Italy, the climate crisis is very real. They regularly face inundation problems during high tides or storms. For other communities, citizens have urged leaders to take on the climate challenge not because it is directly impacting them at the present moment but because it is the right thing to do. This chapter will explore a range of issues associated with climate change in cities, towns, and communities and what we can do about it. The chapter will start by exploring several examples of what some cities have done to address climate change. As we will see, there is not one standard approach and there are considerable regional differences. The chapter will then break down the types of things that community governments control in order to better understand where local leaders have power to enact change. Finally, the chapter will highlight projects that any community can take on within the framework of simple initiatives, hard or expensive initiatives, and innovative and/or life-changing initiatives.

It is important to note that this chapter mixes the power of government with the power of citizenry and activists. It is often the activists that drive the government to do the hard climate work in communities. However, community groups often take on important local and regional community projects. The two organizations—local governments and community groups—together are responsible for the majority of local climate initiatives in cities around the world. As already noted, there are distinct differences based on geography. However, there are also differences based on the nature of local governments, the types of groups active in the community, the key concerns of communities, and local leadership. As a result, there is a fascinating array of differences in how communities around the world approach the climate crisis.

Examples of Communities that Developed Climate Change Policies

Many local governments and communities have developed their own unique approaches to addressing climate change. As we will see, some have taken on particular projects that address one component of the issue while others are comprehensive in trying to achieve carbon neutrality. This section will focus on two cities to showcase the differences: Louisville, Kentucky, and Cairo, Egypt.

Louisville, Kentucky, is a city in the east-central portion of the United States and it has a population of just over 600,000 people. The broader metro region has a population of about 1.2 million. While Kentucky is widely known as the heart of America's coal industry, the city of Louisville has become known for its unique approaches to managing climate change. The city, starting in 2012, under the leadership of Mayor Greg Fischer, embedded sustainability as a key theme in city operations. Today, sustainability efforts are managed within the city's office of planning.

The goals of Louisville's first sustainability plan, which was published in 2012, are listed in Table 6.1 (LouisvilleKy.gov, 2020). As can be seen, Louisville developed a very comprehensive and aggressive plan that sought to focus on several areas of sustainability—particularly climate

Table 6.1 Goals of the Louisville, Kentucky, sustainability plan (LouisvilleKY.gov, 2020)

Focus area	Goals	Target date
1. Energy	1. Decrease energy use citywide per capita by 25%	2025
	2. Decrease energy use in city-owned buildings by 30%	2018
2. Environment	3. Mitigate the risk of climate-change impacts	2018
	4. Achieve and exceed National Ambient Air Quality standards	Ongoing
	5. Improve waterway quality	2018
	6. Increase recycling citywide by 25%	2015
	7. Achieve 90% residential recycling participation	2025
	8. Divert 50% of solid waste away from the landfill by 2025 and 90% by 2042	2025
3. Transportation	9. Decrease transportation-related greenhouse gas emissions by 20%	2020
	10. Reduce vehicle miles traveled by 2025	2025
4. Economy	11. Provide opportunities for clean economy organizations and innovators, and develop a qualified workforce to support it	2015
	12. Expand the local food system by 20%	2018
5. Community	13. Increase access to healthy foods by 20%	2018
	14. Increase opportunities for active living	2015
	15. Incorporate sustainability into the Land Development Code and the Comprehensive Plan	2015
	16. Replace and reforest parks property and provide nature-based recreation	2018
	17. Expand green infrastructure incentives citywide	2018
	18. Establish a robust urban tree canopy and implement strategies to mitigate the urban heat island effect	2018
6. Engagement	19. Engage the community in sustainability practices and principles	Ongoing

change. After the report was published, the city produced a series of annual reports to communicate its progress to stakeholders. The city made tremendous strides in reducing greenhouse gases (GHGs) and in a range of other goals. Unlike other cities that developed sustainability plans to have them languish due to a lack of funding or personnel, Louisville made great progress. For example, by 2017 the city achieved a 10% reduction in greenhouse gas emissions.

While Louisville has a comprehensive plan that addresses a range of issues from food to energy, the city has specialized to a certain extent on its urban heat island. Urban heat islands occur in cities because of the lack of natural surface cover. Heat is absorbed by concrete, asphalt, and buildings thereby causing temperatures to rise. In some cities, like Louisville, the contrasts between the city and its surrounding area can be quite high. Some parts of Louisville, for example, have temperatures that can be 10 °F higher than the surrounding area. These high temperatures can cause health problems and can lead to greater energy use to try to cool homes and offices.

A study of Louisville's heat island, published in 2016, provides one of the most comprehensive studies of its kind ever completed in a city. It highlights the problem, the variability of the heat island across the city, the vulnerability of different areas, and heat management strategies. The city has taken the study seriously and rapidly reacted to its results (Louisville Open Data, 2020). Since 2016, the city has planted tens of thousands of trees, modified roofs to reflect heat back into the atmosphere, and planted vacant properties with grasses. They have also done a tremendous amount of community education to educate the public about heat islands and why it is important to maintain an urban tree canopy to mitigate the impacts of urban heating. Mitigating urban heat islands is a unique way to confront climate change that involves storing carbon in trees while finding ways to reduce energy use by lowering ambient temperatures across the urban landscape.

The Louisville example demonstrates that cities can not only have comprehensive sustainability plans that include climate change, but also that they can take on unique climate change projects that are appropriate for their region. As we will see in the next example, specialization is often imposed by crisis—and often imposed by external forces.

Cairo, Egypt, is one of the most populated regions of the world with a metropolitan population of over 20 million. The city has many environmental problems, but one of the most significant ones is rampant air pollution (Fig. 6.1). The region is also vulnerable to climate change.

6 How Your Community Can Reduce Its Greenhouse Gas Impact

Fig. 6.1 Egypt has notoriously problematic air pollution problem caused by a variety of emissions including car exhaust and local cooking fires

Rising sea levels are already causing problems in some areas of the Nile delta region (Sharaan & Udo, 2020). The Nile, which is Egypt's lifeline, is expected to be stretched thin as the climate gets warmer and drier in the headwaters and as greater demands are placed on it from countries like Ethiopia and Sudan. The coming decades could be quite challenging ones for the entire country.

But Cairo's air pollution problems are immediate and have reached crisis levels. The main challenges in Cairo are different from those in Louisville and they have been addressed rather differently. In 2008, Egypt published a plan called Cairo 2050. It focused heavily on making Cairo a modern global city and included elements of environment sustainability. Since the revolution of 2011, some progress has been made on the plan, but there remains a myriad of environmental challenges in the region. One of the most problematic of them is air pollution which is driven largely by open burning of garbage and the transportation sector.

Unfortunately, scarce government resources have not been used to advance the Cairo 2050 plan since the revolution to significantly address the problem.

The World Bank announced a new $200 million loan to support the Greater Cairo Air Pollution Management and Climate Change Project which focuses heavily on trying to reduce the Cairo region's significant air pollution issues (World Bank, 2020). A 2019 World Bank study demonstrated that particulate matter in the air is a major driver of air pollution (World Bank, 2019). The pollution is so bad that it causes elevated levels of heart disease, stroke, and lung cancer. The study also notes that the greatest sources of pollution are open burning and air pollution from traffic.

The World Bank loan is in support of projects to try to not only reduce the air pollution problem in Cairo, but also to reduce greenhouse gas emissions to support international efforts to address climate change. This example demonstrates that initiatives to reduce greenhouse gases also have benefits in reducing ambient air pollution that have direct implications on public health. By taking steps to reduce greenhouse gas pollution, Cairo is directly improving the health of millions of its citizens.

Regional Greenhouse Gas Inventories

For the last few decades communities have developed a range of options to try to address greenhouse gas pollution. For the most part, there is not one uniform approach. As was seen in the prior examples, cities like Louisville and Cairo developed their own path to manage climate change that worked for them. Large geographic and cultural differences between cities suggest that this is in some ways appropriate. However, there are some basic methods that can be used to assess how to approach greenhouse gas reductions. These methods involve ideas that were introduced in the previous chapter on personal approaches to greenhouse gas reduction that involve conducting a greenhouse gas inventory and following that up with reduction targets and plans to reach them.

The basic idea of a personal greenhouse gas inventory is relatively simple. However, how does one conduct a greenhouse gas inventory in a city

6 How Your Community Can Reduce Its Greenhouse Gas Impact   109

or region? There are many sources of greenhouse gases and it is difficult to capture them all within an inventory. Plus, what types of emissions do you measure? Are all emissions counted, no matter who produces them, or do governments only assess what they are responsible for producing? As you can imagine, a great deal of thought has gone into this issue over the last few decades and some distinct methods have emerged. This chapter focuses on initiatives to measure greenhouse gases produced by government operations. The next chapter takes on the challenging topic of greenhouse gases produced by businesses, non-profits, and schools.

While there are a range of greenhouse gas inventory tools available, there are two initiatives that are important to review related to greenhouse gas inventories for communities. The first is an international inventory tool that was developed by ICLEI and the other is one developed by the United States Environmental Protection Agency.

ICLEI and International Community-Based Greenhouse Gas Inventories. Many people are not familiar with ICLEI, but it is an international organization that emerged out of the United Nations to help local communities work on local sustainability projects. Once part of the United Nations, ICLEI is now an independent non-profit that works internationally to advance the original goals of enhancing local sustainability. It was founded in 1990, an important year in the development of the modern environmental movement. This is the year the European Environmental Agency was established and it was the year of Redwood Summer, a major activist intervention in support of California's old growth forest. Also, the Intergovernmental Panel on Climate Change published their first report. It was a time of growing environmental awareness during an era of accelerating globalization. The world was not only connecting economically through loosening trade barriers, but it was also connecting through emerging understanding of global planetary crises like global climate change.

When ICLEI started, the idea of greenhouse gas inventories was relatively new. However, over the years, they developed international expertise on not only how to conduct greenhouse gas inventories, but also how to follow up with interventions to make the needed reductions. ICLEI has developed three main protocols for managing greenhouse gases within a local context: the US Community Protocol for Accounting and

Reporting of Greenhouse Gases, Global Protocol for Community-Scale GHG Emissions, and the Local Government Operations Protocol. Each of these protocols follows more or less the same process (ICLEI, 2020).

The EPA and Greenhouse Gas Inventories. For many years, the US federal government did not provide any regulation of greenhouse gases as leaders in the executive branch of the government did not direct the EPA to provide any national guidance on climate change pollutants. As a result of the lack of leadership on this issue, a number of lawsuits moved forward to try to force the EPA into recognizing that greenhouse gases produced significant problems for the United States and that they needed some degree of regulation. The lawsuits culminated in a 2007 ruling of Massachusetts et al. vs. the EPA that essentially ruled that the EPA must regulate greenhouse gases because it is clear that there is property damage that can be directly traced to climate change (US Supreme Court, 2007).

What is interesting about this case is that prior to the ruling, the EPA and other US agencies were providing some guidance to the public by providing information on a variety of climate change impacts and mitigations. For example, the EPA did develop a number of greenhouse gas inventory tools for local governments (EPA, 2020a) and also conducted national inventories of greenhouse gases dating back to 1997 (EPA, 2020b). What this means is that the US government did recognize that greenhouse gases were pollutants, but prior to 2007, the government had no interest in regulating them at the federal level. Indeed, the EPA's efforts, to date, have been relatively weak and much of the heavy lifting in reducing greenhouse gases is done by local governments, by non-profit institutions like schools, and by businesses. As will be discussed in more detail later, conducting greenhouse gas inventories, especially measuring Scope 3 emissions, is very challenging. That is why many argue that local inventories are the appropriate scale for assessing how to reduce greenhouse gases.

Conducting Greenhouse Gas Inventories at the Community Level. Due to the complexity of the ICLEI and EPA procedures, it is impossible to review them completely in this space. However, a summary of some of the main activities follows. It is worth noting that both organizations provide many of the resources for conducting greenhouse gas inventories for free. However, ICLEI does offer services that are fee-based and one

6 How Your Community Can Reduce Its Greenhouse Gas Impact

can hire a consulting group to conduct community-based inventories if there is not the bandwidth within an organization to take on the challenging work of conducting one.

Prior to reviewing community-based greenhouse gas inventory procedures, it is worth reviewing the three main sources of greenhouse gases, Scope 1, Scope 2, and Scope 3, gases that were introduced in the previous chapter (Table 6.2). All community-based greenhouse gas inventories divide their efforts into collecting information to assess emissions within each scope. Scope 1 and 2 emissions are probably the easiest to assess because the data are readily available. Scope 1 emissions include those emissions that are produced directly from power generation on site. Thus, if a community produces energy from a small power plant, the amount of greenhouse gases released from the production of energy in that plant

Table 6.2 Definitions of Scope 1, Scope 2, and Scope 3 emissions

Type of emissions	Description	Types
Scope 1 emissions	These are emissions produced directly by an organization. In most cases, these emissions are produced by local energy production or from energy produced by fleet vehicles.	Local on-site power plant Emissions from fleet vehicles
Scope 2 emissions	These are emissions produced from the purchase of energy. For example, Scope 2 emissions would include emissions produced from the emissions of a power plant from which a community purchases electricity.	Emissions produced in the production of purchased electricity or other power sources
Scope 3 emissions	These are emissions produced by entities not owned or managed by an organization.	Emissions produced from employee commuting or travel Emissions produced through the production of waste Emissions produced through wastewater and sewage Emissions produced as a result of purchasing

would be included in the Scope 1 inventory. Scope 2 emissions are also relatively easy to calculate because they are calculated based on the amount of energy purchased. Because most communities purchase energy from one or two sources—say an electric power company and a natural gas company—these Scope 2 emissions can be calculated using annual energy bills.

Scope 3 emissions are the most challenging to calculate because they are the amount of emissions produced by an organization due to their overall activities not associated with energy production. They include things like employee commuting, company travel, energy associated with purchasing, emissions associated with sewage and wastewater, emissions associated with agriculture production, and emissions associated with garbage and special events (Fig. 6.2). As can be imagined, it is hard to get data on all of these categories.

Fig. 6.2 Every community can calculate their greenhouse gas emissions including those associated with festivals like this Oyster Festival in Oyster Bay, New York

Based on the processes identified by ICLEI for Local Governments and the US EPA, community greenhouse gas inventories can be divided into five stages (Table 6.3): (1) setting boundaries, (2) defining scopes and sources, (3) calculating emissions, (4) setting goals and tracking emissions, and (5) improving the inventory and verifying results.

Setting Boundaries. The first stage of a greenhouse gas inventory, setting boundaries, is key to framing how to conduct the greenhouse gas inventory. In some cases, the geographic boundaries are simple, for example, the confines of a single city. In such situations, there is a single chain of command for undertaking the inventory and the boundaries are well-defined. Often, in such settings, there are only a handful of Scope 1 and 2 sources which makes conducting the inventory relatively simple. However, in other cases, setting the geographic boundary can be much more complex and trickier. For example, imagine a situation where a county seeks to conduct a greenhouse gas inventory. Most counties contain a number of other local governments like towns and cities. Does the county inventory include all the local governments or only the parts of the county outside of the jurisdiction of the towns and cities within the local government?

The boundary setting for a community greenhouse gas inventory also includes decisions over what parts of the community are included in a greenhouse gas inventory. Does the inventory only include the actions of the government? If so, these types of inventories must take into account

Table 6.3 Major steps in a community-based greenhouse gas inventory

Step	Action
1. Setting the boundary	1. A. Defining the geographic boundary of the study. Is it a single government entity, or is it a regional study?
	1. B. What is the time frame of the inventory?
2. Defining the scopes and sources	2. A. What are the emission scopes (Scopes 1, 2, and 3) for the study?
	2. B. What data are available for each of the defined scopes?
3. Calculating emissions	3. A. Collect data.
	3. B. Calculate the greenhouse gas emissions for all sources identified in Step 2.
4. Set goals and track emissions	4. A. Set goals for reduction in each scope area.
	4. B. Track emissions over time to assess goals.
5. Improve inventory and verification	5. A. Enhance inventory as data collection and methods improve.
	5. B. Verify inventory with third party.

all government operations and the actions of all employees. Or, does the inventory take into account ALL actions within the county that would include not only the actions of government but the actions of all citizens, businesses, and organizations within the community boundaries? Most community greenhouse gas inventories focus on the actions of local governments given the difficulty of calculating the actions of individuals and local businesses.

Another important issue associated with boundary setting is the time frame. Greenhouse gas inventories must be conducted over a particular period of time. Most, but not all, inventories are done on the annual basis. Such inventories usually capture the ups and downs of annual energy usage and other patterns such as commuting. However, some inventories are done over longer or shorter time frames. Regardless of approach, the initial time period becomes the period to which all other future inventories are compared. Since the overall goal is to reduce greenhouse gas emissions, the initial inventory is important because it becomes the starting point for developing reduction goals.

Defining Scopes and Sources. Community greenhouse gas inventories include an assessment of scopes and sources. What is meant here is that prior to calculating the emissions releases in a community, an inventory of the types of sources within each scope must be conducted. For example, a community must identify all Scope 1 sources by ascertaining how many power-generating facilities are in the community. It must also be able to find the data on energy production at these facilities to calculate emissions. Likewise, a community must be able to find data on the types and amounts of energy used in the community to calculate Scope 2 emissions.

Because Scope 3 emissions are so variable, and are often the major source of community emissions, a great deal of attention is given to assessing what sources exist within the community and how to calculate them. There are a variety of tools for assessing areas of Scope 3 emissions from areas such as wastewater and garbage that take into account population size, nature of wastewater, or waste management technology. This phase of the greenhouse gas inventory must select the assessment tools for Scope 3 emissions and collect the data that is available to conduct the inventory.

Calculating Emissions. Once a community identifies the boundaries and scopes and identifies data sources, the next phase of the inventory is the actual calculation of emissions. As noted earlier, there are a number of tools available for completing this part of the inventory. ICLEI and the US EPA, and many other organizations, provide spreadsheet tools that allow a community to enter basic data on Scope 1, Scope 2, and Scope 3 emissions to get a total for each scope as well as an overall total of the community's emissions. Each of these totals from an initial inventory set the standard from which strategies are developed to reduce emissions.

Setting Goals and Tracking Emissions. Once a community has completed the calculations for each scope, it can set goals for emission reduction and begin tracking emissions over time. For example, a community could set a reduction target of cutting emissions by 10% from each of the scopes. In doing so, it would identify a range of strategies it could employ to reach the goals and begin tracking emissions regularly over time. To reduce Scope 1 emissions, it could move toward producing its own green energy by investing in solar or wind generation. To reduce Scope 2 emissions, it could reduce energy demand by putting in energy-efficient equipment. Because Scope 3 emission sources are complex, a range of actions would likely be undertaken such as finding ways to reduce employee commuting, reducing garbage production, find enhancing wastewater technology such as turning waste into energy through sewage digestion. As these activities go forward, emissions tracking will help to assess if community greenhouse gas reduction goals are being met.

Improve Inventory and Verification. The final stage of a greenhouse gas inventory is assessing the overall process and seeking some type of verification. Greenhouse gas inventories are complex and thus problems can emerge as they are conducted. For example, some communities find it necessary to estimate emissions from employee commuting. However, with time, it may be possible to refine the inventory by geocoding the home locations of employees to better calculate their commute. It may also be possible to get detailed information on their vehicles to further enhance the inventory. As time goes on, many communities find themselves improving on their inventory methodology and accuracy.

Many communities also have their inventories verified by third parties to ensure that what they are doing is accurate and falls within the best practices available for conducting them. A number of organizations, like ICLEI for Local Governments, will help a community with greenhouse gas emissions verifications. In addition, many consulting firms specialize in conducting greenhouse gas inventories. Regardless of approach, employees of the community will have to work with the third party to provide access to information and administrative assistance.

As can be seen in the description of the process, a greenhouse gas inventory is complicated and involves not only measuring greenhouse gases, but also developing a strategy for how they can be reduced. Plus, a greenhouse gas inventory is not something that a community does once and then moves on to other issues. They have to be conducted on a regular basis and someone in the organization needs to stay on top of emerging methods and opportunities for assessing how best to measure and mitigate greenhouse gases produced in the community.

Communities have found a range of ways to manage greenhouse gases. Sometimes, they are managed in-house with existing staff in planning or environmental departments. This approach is not always the best because staff are pulled away from existing duties to take on a greenhouse gas inventory job that supervisors think may be simple. Plus, existing staff may not have the skills to undertake the inventory. Another approach is to put together a committee of individuals with a range of talent across local government. This approach is particularly effective in getting access to the range of data that is needed to conduct a greenhouse gas inventory. Another approach is to hire new staff with a title like Sustainability Coordinator to undertake the inventory. These types of positions can be effective if they are truly empowered to gain access to information across a range of departments in local government. However, they can be set up for failure if they are not given the appropriate authority to work with department heads across the local government space. A final approach is to hire a consulting firm to conduct the inventory. The downside of this approach is that the consulting firm does not usually engage with the community to continue to refine and update inventories and mitigation plans.

What seems to really matter in many cases in local government is leadership. When there is a strong leader that makes sustainability a priority, local government usually reacts effectively in developing the needed inventories and mitigation plans regardless of how the inventories are managed. All across the world, one can find examples of how local leadership of government employees or citizen stakeholders are driving change by conducting greenhouse gas inventories and developing projects to reduce the impacts of local government on our climate.

Key Government Initiatives

Local governments have tremendous opportunities to reduce or even eliminate their overall greenhouse gas emissions by transforming their energy and transportation infrastructure, the two biggest producers of greenhouse gas emissions. These are reviewed in Table 6.4. In terms of energy use, there are a range of small steps that can be taken such as converting oil or coal power systems to natural gas or electric. Plus, a community could start to purchase carbon credits that would mitigate greenhouse gas emissions. More difficult or more expensive choices include developing protocols for requiring solar on homes and developing local green energy projects. Plus, one of the most radical approaches would be to commit the community to become a zero-emissions community that had no greenhouse gas emissions as a result of energy use.

The transportation sector in a community can also be transformed. Easy efforts to mitigate greenhouse gases would be to change community fleet to hybrid or electric vehicles and to encourage online meetings over travel to conferences or to meet stakeholders. Harder or more expensive initiatives would be to develop mass transit options and promote walkable communities. Communities could also encourage employees to work from home to avoid commutes. More innovative approaches would be to reduce street sizes and parking availability. Some communities are building new infrastructure around golf carts or bikes to reduce the need for road infrastructure and others are planning around shared driverless cars to reduce the need for the number of vehicles in the community.

A community's building infrastructure can also be enhanced to reduce greenhouse gases through zoning and the development of new infrastructure. One of the easiest things a community can do is to enact green development regulations and to change zoning rules to promote

Table 6.4 Simple, hard/expensive, and innovative/life-changing approaches to reducing greenhouse gas emissions in a community from energy production, transportation, and buildings

	Energy	Transportation	Building and community infrastructure
Simple	Purchase green energy certificates Convert any petroleum or coal-based energy systems to natural gas or electric	Convert community fleet to hybrid or electric vehicles Encourage online meetings over travel to meetings or conferences	Promote densification projects Develop regulations for green building
Hard/expensive	Change zoning laws to require solar on all homes Develop economic incentives for the development of green energy in the community	Provide infrastructure for electric cars Prioritize planning to promote walkability, mass transit, and carless living Promote working from home when possible	Reduce public building space by developing work from home policies Develop waste to energy infrastructure for garbage and sewage
Innovative/life-changing	Commit to zero emissions in the community from energy Aggressively pursue a range of green energy initiatives	Reduce street size and parking Advance driverless car technology Reduce car infrastructure with golf cart or alternative vehicle infrastructure Ban travel reimbursements for employees	Commit to build all public buildings to US Green Building Council's gold or platinum standards Develop zoning rules to allow tiny houses and cohousing

densification in important areas. Densification enhances mass transit and also creates energy efficiencies. Harder or more expensive options are to change work policies so more workers can work from home. This reduces the need for public building space. Also, communities could build waste to energy infrastructure to deal with garbage and sewage. Some of the more innovative actions that could be taken include the requirement to build all public buildings (or even any buildings) to the US Green

Building Council's gold or platinum standards. In addition, communities could change regulations to promote tiny home living and cohousing.

There are many other unique options that communities could take to reduce greenhouse gases. For example, they could follow Louisville, Kentucky's lead by planting trees to promote shade and also carbon storage. Or, they could advance new smart innovative energy management systems that provide power to buildings only when needed. Regardless of approach, communities all over our planet are changing the way that they manage themselves to reduce greenhouse gases. It all starts by conducting a greenhouse gas inventory and finding out where there are opportunities to reduce emissions.

References

EPA. (2020a). Local greenhouse gas inventory tool. Retrieved December 5, 2020, from https://www.epa.gov/statelocalenergy/local-greenhouse-gas-inventory-tool

EPA. (2020b). U.S. greenhouse gas reporting archive. Retrieved December 5, 2020, from https://www.epa.gov/ghgemissions/us-greenhouse-gas-inventory-report-archive

ICLEI. (2020). Greenhouse gas protocols. Retrieved December 5, 2020, from https://icleiusa.org/ghg-protocols/

LouisvilleKy.gov. 2020. Sustainability, Retrieved December 5, 2020 from https://louisvilleky.gov/government/sustainability

Louisville Open Data. (2020). The urban heat island project. Retrieved December 5, 2020, from https://data.louisvilleky.gov/story/urban-heat-island-project

US Supreme Court. (2007). Massachusetts vs. EPA. Retrieved December 5, 2020, from https://supreme.justia.com/cases/federal/us/549/497/

World Bank. (2019). Concept project information document (PID)—Egypt: Greater Cairo air pollution management and climate change project—P172548 (English). Retrieved December 5, 2020, from https://documents.worldbank.org/en/publication/documents-reports/documentdetail/622831576851407904/concept-project-information-document-pid-egypt-greater-cairo-air-pollution-management-and-climate-change-project-p172548

World Bank. (2020). Greater Cairo air pollution management. Retrieved December 5, 2020, from https://projects.worldbank.org/en/projects-operations/project-detail/P172548?lang=en

7

How Your School, Non-profit Organization, or Business Can Reduce or Eliminate Its Carbon Footprint

Introduction

Over the last few decades, schools, non-profit organizations, and businesses (SNBs) have made significant strides in finding ways to reduce or even eliminate their greenhouse gas emissions. What makes these organizations so interesting to examine is that each of them have very different missions and approach greenhouse gases in different ways. Schools, for example, often utilize the reduction of greenhouse gases as a way to educate their stakeholders on a range of climate change issues. Non-profit organizations typically frame their efforts within the realm of community improvement. In contrast, businesses focus on greenhouse gas reduction as a sound way to ensure the continuation of their activities. An unhealthy planet is not particularly good for most business enterprises.

This chapter takes a look at what SNBs have been doing to reduce greenhouse gases. The section that follows takes a look at what some K-12 schools and institutions of higher learning have accomplished. As we will see, some of the efforts have far-reaching impacts that help to transform society to have a deeper understanding of climate change.

Following the section on educational institutions, the chapter moves into an examination of the efforts of non-profit organizations to see how they are changing their activities. Next, the chapter delves into the efforts of the business world. Many companies have made tremendous strides in changing their business models to include climate-friendly policies.

It is worth noting that many of these organizations also conduct detailed greenhouse gas inventories to understand the major sources of emissions. Because the methods of greenhouse gas accounting were covered in detail in the previous chapter, they will not be reviewed again here. Suffice it to say that it is important for all of these organizations to consider their greenhouse gas contributions in order to understand how to make wise reduction decisions. This chapter, however, focuses on key initiatives that address an organization's overall greenhouse gas emissions without taking into consideration the inventory process. Please note that it is highly recommended that the starting point for any organization is the greenhouse gas inventory.

Schools

Schools around the world vary considerably. In one way, they can be classified by the age of the students. For example, schools can focus on education of young children at elementary schools or adult education at colleges at universities. They can also be classified by size. Some schools are rather small in rural areas while some universities can have tens of thousands of students. Some schools have elaborate buildings and technology while others are simple and may even be outdoor. Regardless of size or age, the overall purpose of all schools is education. They provide a distinct curriculum to help students learn. Some schools are public schools that have a set curriculum while others are private schools which often have a curriculum that is influenced by a particular mission set by their boards or leadership teams.

Schools also have other purposes. They help to create educated citizens that drive social good and they also advance economic development goals by creating an educated workforce. There is no doubt that societies that put a strong emphasis on education thrive more than those that do not. Schools help to create social cohesion by providing a uniform approach to education. Many have argued that a strong public education is key to maintaining strong democratic traditions around the world. Where there are weak educational approaches, democracy suffers and fascism and totalitarianism thrive. In such situations, decisions are often made that are not in the greatest public interest because the public is not in a position to intellectually question leaders.

Local and regional economies are also dependent on an educated workforce. While basic skills such as reading and math are more and more universally taught around the world, schools also advance specialized learning. Complex economies thrive in places where there are advanced centers of learning that encompass not only higher education at the university level, but also vocational training for adults and strong elementary and secondary education opportunities for children. More and more, educational institutions are embracing a range of curricula and policies that advance greater understanding of climate change and how to mitigate its impacts on our planet.

Sustainability Initiatives in K-12 Schools. The range of options available to K-12 schools is listed in Table 7.1. At the most basic level, sustainability can be incorporated within the curriculum. While many schools have lockstep curricula, some schools have found ways to easily add units on sustainability, particularly when using the school's facilities to teach about specific issues of sustainability. For example, a school located near a water system such as a pond or river can utilize the water feature to teach a unit on pollution or nutrient cycling that leads to broader conversations around global climate change and the impact of greenhouse gases in the environment. Some schools have also invested in green energy projects like solar or wind on their grounds to teach young people about energy use and production, green energy, and climate change. Many schools have also built school gardens which is a great stepping off point to discuss the role of agriculture in climate change—particularly the impact of meat production. Some schools have also completely transformed their mission to focus exclusively on sustainability and are seeking to make transformative change in their region.

A great example of this type of school is Learning Gate Community School, a K-8 school in Tampa, Florida (Learning Gate Community School, 2020). This school has a distinct mission and vision that clearly states its role in sustainability education:

Table 7.1 Simple, hard/expensive, and innovative/life-changing approaches to sustainability in schools

	K-12	Higher education
Simple	Infuse sustainability within the curriculum Use facilities to teach transformation	Require sustainability in general education courses Create a major in green energy or sustainability Join AASHE
Hard/expensive	Create school gardens Install solar energy systems	Create a sustainability office Conduct a greenhouse gas inventory Conduct a STARS assessment Pay for carbon credits for all travel
Innovative/life-changing	Make sustainability a key mission of the school	Divest from fossil fuels All vegan campus food options Create a carbon neutral campus

Our Mission: To promote academic excellence, community service and environmental sustainability through family and community partnerships.

Our Vision: Tomorrow's leaders engaging in and contributing to an educated, sustainable world.

Learning Gate has educated thousands of students within this clear framework. Mission-oriented schools have the ability to help with social transformation. Getting support for this type of mission from parents and the broader community can often be challenging due to the politicization of sustainability and climate change in our global culture.

There are many educational leaders who are actively working to help schools with the transition to sustainability. Steven Ritz, for example, started his path toward a global sustainability education leader by teaching his students a range of sustainability topics within the realm of food (Green Bronx Machine, 2020). His at-risk students saw significant improvements in a variety of performance indicators including grades, attendance, and behavior. Plus, because he focused on growing healthy food, he was able to educate students about healthy diets and provide fresh vegetables to students and their families. Since his early days of teaching in the South Bronx, Ritz has branched out to focus on a variety of sustainability issues around the world.

Ritz is but one example of how one educational leader can make a difference. All over the world, teachers with a passion for the environment and sustainability are transforming schools and students' lives.

Sustainability Initiatives in Higher Education. Table 7.1 lists some of the key initiatives that schools can undertake to drive forward a climate-positive agenda. Higher education in most places is very different from K-12 systems in that there is much more local control of the curriculum. Thus, education at colleges and universities varies considerably and often reflects regional economic differences, faculty expertise, and the mission of the institution. For example, an agricultural school in Australia will have a very different curriculum from a vocational college in Paris. Thus, there is room within the curriculum to advance sustainability education using a local, regional, or national lens. Some schools have created majors in green energy or sustainability while others have advanced sustainability courses within their general education curriculum.

In the United States, an organization called the Association for the Advancement of Sustainability in Higher Education (AASHE) helps higher education institutions reach their sustainability goals (AASHE 2020a). They provide member institutions with a range of opportunities including an annual conference, best-practice resources, and some tools for conducting campus assessments. As of late, AASHE has been providing memberships for non-US institutions. However, their main focus is on North America. When an institution joins AASHE, all members of the institution have access to a range of resources that encompass the breadth of university operations. Thus, faculty, food service specialists, grounds keepers, facilities managers, administrators, and others will all find some information that will be useful within AASHE's umbrella.

Some colleges and universities have opted to create an office of sustainability that manages and coordinates an institution's sustainability undertakings. These positions are sometimes located within the academic side of the house and sometimes within facilities. Each setting has its plusses and minuses. When a sustainability officer is located within academic affairs, the focus is typically in the areas of curriculum and student engagement. These positions typically have little authority or influence over facilities management. Thus, when the positions are located within facilities, they can significantly impact a range of university facilities operations like energy, food, and water use. Some schools have gotten over this divide by having the sustainability officer manage a sustainability committee that brings together a range of stakeholders and that is empowered to make real change.

Some universities also undertake greenhouse gas inventories and conduct STARS assessments (AASHE 2020b). The STARS program is a sustainability assessment tool that allows colleges and universities to conduct a self-assessment to measure their sustainability. The assessment rating system, like the US Green Building Council's rating systems, rates institutions within bronze, silver, gold, and platinum categories. STARS assessments are major efforts and require the buy in of numerous offices of an institution. It also requires some expertise in measuring and assessing sustainability within organizations. Some universities have charged committees with completing STARS assessments, while others have made one individual or office in charge of the project. Regardless of approach, it does require quite a bit of coordination and the assessment is not an easy undertaking.

Another challenging effort that some colleges and universities have taken on is ensuring that all travel is carbon neutral. Because university faculty and students travel quite a bit to conferences or for research projects, the travel component of a greenhouse gas inventory for higher education tends to be relatively high compared to other organizations. As a result, some universities have developed policies that require the purchase of carbon credits when making flight arrangements. Others have purchased large values of carbon credits to cover a range of emissions, including travel.

Some of the more challenging initiatives involve some controversial topics. For example, some universities have opted to divest from fossil fuels within their endowment portfolio. Over the last several decades, students and faculty across the world have urged their institutions to invest their endowments within the ethical framework of the organization. Perhaps the most famous and successful divestment program was the effort in the 1970s and 1980s to divest from South Africa due to its shameful apartheid policies. In more recent years, a number of organizations, including 350.org, urged campus activism around divestment to draw attention to the role that large energy companies have in our changing climate (350.org, 2020). In addition, a number of churches and non-profit organizations have moved their endowments from supporting fossil fuels within the aegis of protecting God's creation.

Many universities across the world have divested from fossil fuels including the University of Massachusetts, Stanford University, the University of Bristol, and Oxford University (Fig. 7.1). There is no doubt that the trend will continue (gofossilfree.org, 2020). The effort is bolstered by growing initiatives of investment companies to build investment portfolios that focus on green initiatives. For example, BlackRock, the largest investment firm in the world, announced in 2020 that it would no longer invest in companies that did damage to the environment. Indeed, green investing is a growing international trend in finance.

Fig. 7.1 Oxford University has tried to embrace sustainability in big ways. How have schools in your region embraced sustainability?

Non-profit Organizations

One of the key aspects of non-profit organizations is that they typically are built around a public mission. This book has focused on two very specific types of non-profits: governments and schools. This section focuses on three other types of non-profit organizations: youth groups and clubs, adult service organizations like the Lions Club, and community service organizations that provide a distinct service to particular segments of our communities. While there is some overlap among the groups, and while this selection does not encompass all types of non-profit organizations, it does represent the groups where there is a significant amount of activity on the sustainability front. Due to the vast number of these types of groups, there is a great deal of variability in how they approach sustainability and climate change.

Youth Groups. Table 7.2 lists a range of possible activities that youth groups like scouts, brownies, or others can undertake to address climate

Table 7.2 Simple, hard/expensive, and innovative/life-changing approaches to reducing greenhouse gas emissions in three different types of non-profits: youth groups, adult service groups, and community service organizations

	Youth groups	Adult service organizations	Community service organizations
Simple	Provide training or service opportunities for youth	Member education Provide opportunities for adult community service around climate change	Frame organizational operations to become carbon neutral
Hard/expensive	Provide experiential opportunities for youth to experience the impacts of climate change	Organize community support around key climate change projects in the community	Educate stakeholders about the intersection of the organization's mission with climate change
Innovative/life-changing	Engage youth in the Fridays for Future initiatives	Reorient the organization's mission around the urgency of climate change	Reorient an organization's mission to include issues of climate change and climate justice Sign and commit to the NGO Climate Compact

change. At the simplest level, groups can provide ways to infuse sustainability within youth training. For example, to earn a Sustainability Merit Badge in the Boy Scouts of America, members must be able to understand a variety of sustainability issues including climate change and energy conservation. At a deeper level, some groups are providing opportunities to experience the impact of climate change on communities. For example, many church youth groups do mission-related work in places that have been heavily impacted by climate change. For groups that want to delve even deeper into more challenging conversations, they could teach their members about Fridays for Future or the climate strike movement. Greta Thunberg is a major figure among young people and provides a beacon of hope for many (Associated Press, 2019). The activities associated with climate strikes have inspired young people and adults all over the world to protect our planet and reduce the impacts of climate change.

Adult Service Organizations. Around the world, there are many different types of adult clubs and service organizations. Some of the largest are organizations like The Shriners Club, The Lions Club, The National Organization of Women, and The Masons. While each of these clubs has distinct and well thought out missions, there is a great deal of intersectionality between climate change and many of their key concerns. Table 7.2 lists some of the areas where these organizations can make a difference in the realm of climate change.

At the most basic level, it is worth considering how members could be educated on climate change and how they can get involved with service projects that address some component of climate change in their community. Some of these opportunities might be relatively clear such as doing tree planting or assisting with mitigating a flood disaster. However, some may be intersectional and opportunities for a deeper understanding of climate change within the context of their mission. For example, an adult service organization working on issues of women may take on projects associated with women impacted by climate change. At a deeper level, organizations often have tremendous community leverage and can build support for community climate change projects. A group focused on enhancing youth entrepreneurialism could help frame regional economic development around sustainability and climate action. A deeper dive for some organizations would involve reorienting their mission to encompass sustainability and climate action.

Community Service Organizations. Community service organizations or non-government organizations that provide public services vary widely. They often have very specific missions such as seeking to address poverty, housing, environmental issues, children's issues, or health. While there are many well-known national or international service organizations, they most commonly have local or regional foci. Table 7.2 provides a list of some of the types of things that these organizations undertake. Because their funding is typically driven by donations or grants, these types of groups sometimes have a difficult time finding resources to put toward climate change initiatives. However, at the most basic level, many have found ways to transform their operations to try to become carbon neutral. Some organizations have taken the more challenging step of educating their stakeholders about the intersectionality between their mission and climate change. For example, if an organization focuses on mental health, the group could seek to communicate how climate change is impacting the mental health of their stakeholders. Many scholars have written about issues of ecological grief and the frustration with our public officials at denying climate change and it is easy to make these connections. Indeed, climate change problems are so challenging across society that it is relatively easy to find connections with most issues that are part of the missions of non-profit organizations. Plus, every region of our planet is seeing impacts in some way and it is easy for local or regional organizations to find links between their work and climate change.

One of the most challenging decisions a group can make is to include climate change within their organization's mission. As the previous paragraph discussed, climate change is disrupting society and making conditions worse for many people. Climate change is driving migration and transforming how places manage housing, employments, agriculture, transportation, and a myriad of other issues. As I write this, half a million people in the US state of Oregon are under evacuation orders due to a massive climate-driven fire. As the climate changes and we see more extremes, there is a need for public service organizations to take a larger role and help address the climate change problems that impact their stakeholders. As a result, some community service organizations have added climate change within their mission or organizational statements. For example, Sustainable Rural Community Development Organization

Malawi has built climate change within a component of their overall goal statement (SRCDOM, 2020). An organization can take even further steps by abiding by the principles of the NGO Climate Commitment as outline in Table 7.3 (Interaction.org, 2020).

Table 7.3 The NGO Climate Compact (taken from https://www.interaction.org/wp-content/uploads/2020/04/Climate-Compact.pdf)

Education and Advocacy
• Improve understanding among our constituents, donors, vendors, corporate partners, and staff of the challenges of climate change, environmental degradation, and loss of biodiversity, and how they connect to development and humanitarian issues;
• Strategically work to make adaptation to and mitigation of climate change public policy priorities within our organization's relevant areas of influence locally, nationally, and internationally;
• Ensure that the voices of affected communities and the local institutions that serve them are represented in policy decisions.
Programs
• Strengthen our organizations' technical expertise in the existing approaches, tools and processes to run climate-aware, environmentally sustainable programs;
• Begin mainstreaming climate and environmental considerations into all stages of programs to reduce our emissions and any environmental degradation resulting from our work; in the process, maximize the resilience and adaptation of affected communities;
• Ensure that the voices from those communities, especially women, youth, and other marginalized groups, are informing and benefiting from that process.
Internal Operations
• Assess the major categories of greenhouse gas emissions, water usage, and waste at our organization's headquarters, country offices, and field offices; consider conducting a baseline study of those aspects using the Greenhouse Gas (GHG) Protocol or other widely accepted industry standards;
• Based on that assessment, propose actions that our organization will take to reduce emissions and waste, and begin implementing them;
• Start engaging our board of directors in an exploration of our organization's relationships to the fossil fuel industry and other industries that generate large-scale negative impacts (e.g., deforestation, pollution) on the environment.
Learning
• Learn about and debate new approaches, tools, and processes in international development and humanitarian work to address climate change and environmental degradation.

Businesses

Our modern global economy helps to drive climate change and many business organizations are rethinking their overall approach in order to reduce or eliminate their overall role in our climate crisis (Fig. 7.2). This is good business for many reasons. Business is likely to be quite bad in a world with climate chaos. Plus, the public is growing dissatisfied with the status quo. In a world where political leaders have failed to effectively address climate change, business leaders are stepping up and having a larger voice in how we manage climate change holistically. Table 7.4 lists a range of actions that businesses can take to address their climate change goals.

At the most basic level, businesses can conduct greenhouse gas inventories of all of their operations. This has become a relatively standard

Fig. 7.2 Ford Motors physically shows their commitment to sustainability in their green roof and with extensive arrays of solar panels. Ford is also transforming many of their car models to electric vehicles

Table 7.4 Simple, hard/expensive, and innovative/life-changing approaches to reducing greenhouse gas emissions in business

	Businesses
Simple	Conduct a greenhouse gas inventory of all operations and make appropriate changes
	Evaluate the need for employee travel and promote work from home
	Communicate changes to stakeholders
Hard/expensive	Evaluate products and services to make them carbon neutral
Innovative/ life-changing	Divest finances from fossil fuels
	Make fighting climate change part of the organizational mission and vision
	Commit to group climate pledge

activity for most large companies and it is becoming much more common among small and medium-sized businesses as well. Greenhouse gas reporting is usually transparent and available for all to see. For example, IBM, one of the largest companies in the world, provides their greenhouse gas inventory information on a website (IBM, 2020). The inventory is a first step in a process of developing clear strategies for reducing greenhouse gases comprehensively in all operations.

Another important step is to evaluate the need for employee travel. The COVID-19 crisis has demonstrated that many business activities that required significant air travel can now be managed using video communications technology like Zoom or Microsoft Teams. In addition, the crisis demonstrated that many operations that required workers in the office can be conducted from home. This eliminates the need for employee commuting and for expansive office space that needs heating and cooling. Some businesses cannot fully eliminate all on-site work or business travel, but it is important to recognize that the COVID-19 crisis has changed the world and we have the opportunity to reexamine some of our activities that are responsible for significant greenhouse gas emissions to see if we really need them.

Another important initial step is to communicate your organization's commitment to addressing the climate problem to your stakeholders. What were the results of your greenhouse gas inventory? What steps are you taking to reduce emissions? Many organizations now have major components of their main websites dedicated to providing this information to the world. For example, the retail giant, Target, has a page dedicated to their greenhouse gas inventory and the commitments they are making to reduce their emissions (Target, 2020). The goals are very specific and they provide links to PDFs of annual climate reports. This level of transparency is becoming the norm in businesses around the world and it is really the first step that many organizations need to take for their commitment to climate change to have any credibility.

The next level of commitment is to transform your products and services to make them carbon neutral. This level of effort can be easy if your product or service is something that can be easily transformed. However, in some cases, the transformation can be difficult if there are some underlying challenges within your business model. An easy example might be a restaurant. Perhaps the restaurant could significantly cut its greenhouse gas emissions by cutting beef from the menu, by focusing on local produce, or by cutting all meat use. While it would not fully eliminate the restaurant's greenhouse gas emissions, it would fundamentally change the organization's impact over time significantly.

An energy company that focuses on petroleum extraction and refinement might be a more challenging example. Yet some companies have found a way to make the transition away from a focus on oil. BP, for example, which used to be known as British Petroleum, states on their website that they "…aim to be a different kind of energy company by 2030 as we scale up investment in low-carbon, focus our oil and gas production, and make headway on reducing emissions. Our new strategy kickstarts a decade of delivery towards our #bpNetZeroambition." BP clearly is aggressively seeking to transition from dirty energy to clean energy within this era of greater awareness around climate change. It is worth noting that most of the major energy companies around the world are shifting rapidly to focus on renewable energy. In the United States, for example, coal energy production has largely been phased out and there has been a tremendous growth in renewable energy in the last two decades. This transformation is taking place all over the world as companies are taking greater responsibility toward reducing greenhouse gas emissions.

Service industries are also transforming. Service industries vary widely and include things like cleaning services, tourism, health care, technology, food, entertainment, and advertising. Each one of these industries has in some way found ways to move toward climate neutrality. Two of these, tourism and technology, will serve as examples. The tourism industry has worked hard in a range of sustainability initiatives over the last few decades and developed a range of initiatives such as ecotourism, service tourism, and green hotels. Leaders in this sector have developed a range of benchmarking tools to create opportunities for large and small vendors to advance green initiatives including greenhouse gas reductions. Green Globe Certification is one of the benchmarking tools available to the travel industry to benchmarking tourism operations. Energy use and greenhouse gas emissions are included as part of the benchmarking and Green Globe provides strategies to assist the tourism industry reach their goals.

As technology use in society has increased, there is growing concern over the impact of technology on greenhouse gas emissions. There are a range of issues that have emerged ranging from the energy needed to mine the rare elements needed to manufacture our devices to the vast amount of energy needed to charge them and produce the wireless networks that help to manage our modern lives.

Some would argue that our technology may not be green itself, but it is helping us become greener. This argument is made because our green technology is helping us run a range of smart technology that improves a range of efficiency metrics that leads to the use of less energy. Smart home technology, for example, helps us heat and cool our homes when needed, turn on lights when needed, and limit the energy consumption on a range of other appliances. While this is true, the technology itself does utilize energy and many around the world are examining how we can cut the greenhouse gas emissions associated with technology.

The IT industry has been examining a range of sustainability issues associated with technology and is framing the issues within the framework of *Green IT*. Green IT initiatives include energy use, but also include things like life cycle assessment, electronic waste, manufacturing of products, and energy efficiency. A recent report from Fujitsu noted that Green IT is a relatively new concept and many in the industry are unaware of the issues (Fujitsu, 2020). Some organizations are taking the issue seriously and are seeking to find ways to reduce the greenhouse gas impacts of technology. Apple, for example, is seeking to be climate neutral in its supply chain and products by 2030 and many tech companies have funded renewable energy production in order to offset the energy use of their servers and energy use of their overall operations (Apple, 2020).

Some of the hardest initiatives that businesses can undertake are divestment from fossil fuels, making climate change and greenhouse gas reduction part of the organization's mission, and committing to regular benchmarking within an industry-wide climate pledge. Some companies invest resources in businesses as part of their business model or as part of a company investment strategy. Some companies have taken the step of not investing their funds in organizations that negatively impact our planet through their actions. Socially conscious investment initiatives have been around for generations. However, in our modern era, they have focused more and more on climate. BlackRock, the largest investment company in the world, is no longer investing in organizations that negatively impact the climate and they join a range of other financial organizations that provide these types of services to green investors.

Many businesses are also reexamining their purpose within our modern climate crisis and have rewritten mission or vision statements to make sure that issues of sustainability and climate are part of what drives their businesses forward. For-profit organizations are understanding that an unstable climate leads to an unstable business community and that they have a responsibility to ensure that they have a role in mitigating their impact. Some companies have gone further and joined other similar organizations within climate or sustainability pledges. Amazon, for example, in 2019 formed a group called The Climate Pledge which seeks to aggressively cut greenhouse gas emissions very rapidly. They seek to be net carbon zero by 2030 and run on 100% renewable energy by 2030. Other organizations, like Verizon, have joined Amazon in the pledge.

What is important to note about all of these areas—schools, non-profits, and businesses—is that they react to public pressure and organizational leadership matters. An organization will only make limited progress if its stakeholders are not interested or if the leadership is not committed to making fundamental change. As a leader in the climate movement, you have the opportunity to move your organization toward climate neutrality.

References

350.org. (2020). *350.org*. Retrieved December 5, 2020, from https://350.org

AASHE. (2020a). The association for the advancement of sustainability in higher education. Retrieved December 5, 2020, from https://www.aashe.org

AASHE. (2020b). The sustainability tracking, assessment & rating system. Retrieved December 5, 2020, from https://stars.aashe.org

Apple. (2020). Apple commits to be 100 percent carbon neutral for its supply chain and products by 2030. Retrieved December 5, 2020, from https://www.apple.com/newsroom/2020/07/apple-commits-to-be-100-percent-carbon-neutral-for-its-supply-chain-and-products-by-2030/

Associated Press. (2019, May 28). Teen climate activist Greta Thunberg addresses leaders at world summit. *The Washington Post*. Retrieved June 1, 2019, from https://www.washingtonpost.com/lifestyle/kidspost/teen-climate-activist-greta-thunberg-addresses-leaders-at-world-summit/2019/05/28/a34c7d04-7cb3-11e9-8ede-f4abf521ef17_story.html?utm_term=.e327a3aa439b

Fujitsu. (2020). Green IT: The global benchmark. Retrieved December 5, 2020, from http://www.ictliteracy.info/rf.pdf/green_IT_global_benchmark.pdf

Gofossilfree.org. (2020). *Gofossilfree.org*. Retrieved December 5, 2020, from https://gofossilfree.org/divestment/commitments/

Green Bronx Machine. (2020). Green Bronx Machine. Retrieved December 5, 2020, from https://greenbronxmachine.org

IBM. (2020). GHG emissions inventory. Retrieved December 5, 2020, from https://www.ibm.com/ibm/environment/climate/ghg.shtml

Interaction.org. (2020). The NGO Climate Compact. Retrieved December 5, 2020, from https://www.interaction.org/wp-content/uploads/2020/04/Climate-Compact.pdf

Learning Gate Community School. (2020). Learning Gate Community School. Retrieved December 5, 2020, from https://learninggate.org/

SRCDOM. (2020). Sustainable rural community development. Retrieved December 5, 2020, from https://www.betterplace.org/en/organisations/8219-sustainable-rural-community-development-surcod

Target. (2020). Climate. Retrieved December 5, 2020, from https://corporate.target.com/corporate-responsibility/planet/climate

Part III

Environmental Sustainability

There are so many environmental challenges that the world is facing that it is sometimes hard to keep up with the latest crisis. We have challenges with energy production, pollution problems, waste issues, and damaged ecosystems. In every corner of the world, these issues seem to be accelerating and leading to lasting environmental change. The following chapters take a look at these systems and provide a deeper understanding of ways that we can take action to try to stop or mitigate the alterations we are all seeing. Chapter 8 focuses on energy by looking at different forms of energy and how we can rapidly move toward greener forms of it. Chapter 9 dives into the vexing problems associated with environmental pollution and how to reduce our responsibility for pollution. Chapter 10 looks into a range of waste issues and Chap. 11 examines the challenging problems associated with ecosystems. As we will see, we have done great damage to our planet in the last century or two and we have much work to do to try to fix the problems we have created.

8

Moving to Green Energy

Introduction

The previous chapters focused heavily on climate change and what you can do in your homes, businesses, and communities to reduce your greenhouse gas impacts. This chapter discusses energy in a different way by highlighting the types and amounts of energy we currently use and the range of options available to us as we move to green energy. In addition, the chapter concludes with distinct suggestions as to how you can be part of the green energy revolution that is underway all over the world.

Dirty Energy Use

The Industrial Revolution that took place over the last few centuries was fueled by dirty energy like coal and petroleum. Three centuries of industrialization has transformed our planet in profound ways via the extraction and transportation of energy sources, through our changing climate due to the pollution associated with the burning of fossil fuels, and via the expanded footprint of humans brought about by the massive amounts of energy used over this time period. The two major dirty energy sources that helped to drive this boom are coal and petroleum.

Coal. Vast reserves of coal in places like Great Britain, Germany, and the eastern United States fueled early industrialization in Europe and North America. In the nineteenth century, coal quickly replaced wood as a driver of industrialization due to its higher energy yield per mass. Plus, coal seemed a limitless resource compared to wood which requires years for regeneration. Another advantage of coal is that it is easily mined and transported. Indeed, early railroads were designed, in part, to move coal from mining sites to sources of labor. Coal use especially expanded in the late nineteenth century when coal began to be used for producing municipal electricity (Our World in Data, 2020a). Thomas Edison built the first coal-burning power plant in London in 1882. Soon after, coal-powered electrical plants dotted the global landscape.

Coal use grew steadily around the world and saw significant increases in the last few decades as industrialization expanded in places in Asia, specifically China and India. While coal use peaked around 2012 and 2013, coal consumption declined around the world by 2.6% between 2018 and 2019 (Enerdata, 2020a). Nevertheless, some countries, especially China, the world's largest user of coal, saw increased use. For a list of the top coal-consuming countries, see Table 8.1. What is interesting in

Table 8.1 Top coal-consuming countries in 2019 (Enerdata, 2020a)

Country	Amount of coal consumed (mT) in 2019
China	3826
India	948
United States	546
Russia	225
South Africa	192
Japan	187
Germany	171
Indonesia	136
South Korea	132
Turkey	121

this table is the massive use of coal in China compared to the rest of the world which is moving very quickly away from coal. The European Union saw one year decreases from 2018 to 2019 of 18% and the United States saw a drop over the same time period of 12%.

The changing coal use really tells two distinct stories. The traditional developed world, most notably the United States and Europe, are moving away from coal due to the issue of pollution and climate change. The energy consumption in these places is not necessarily declining, but the sources of energy are changing to cleaner sources like natural gas, wind, and solar. China, and several other developing countries, are seeing increased use of coal because they are still developing their electrical systems and undergoing an industrial transition. While many developing nations, particularly China, have aggressively worked to develop green energy sources, they are still heavily reliant on dirty energy sources. Plus, many developing nations have vast reserves of coal that provide inexpensive sources of energy.

Petroleum. By the middle of the nineteenth century, petroleum began to make some inroads in coal use as an energy source. Of course, petroleum was a well-known natural material that was used in a variety of products prior to this era. However, with industrialization, new petroleum refining techniques were developed and a range of new innovative products were invented. Early uses of refined petroleum products included using kerosene for home lighting, use of oil for home heating, and use of oils for powering steamships.

With time, petroleum competed with coal for municipal electricity production. Today, globally, petroleum produces 10,000 more TWh of direct energy than coal. Of course, with the advent of the gasoline automobile, new uses for petroleum were born. The top producers of crude oil and the top users of petroleum products are listed in Table 8.2. What is significant about this table is that many of the top consumers are not producers. What this means is that they are heavily reliant on imports for their petroleum energy needs. Crude oil and its refined products are easily transported using a variety of methods such as pipelines, tanker ships, and trucks. However, this transportation system can be easily disrupted

Table 8.2 List of top crude-oil producers and petroleum product consumers in the world as of 2019

Top producers of crude oil	Amount (Mt)	Top consumers of petroleum products	Amount (Mt)
United States	745	United States	760
Russia	560	China	617
Saudi Arabia	545	India	224
Canada	268	Russia	136
Iraq	232	Saudi Arabia	106
China	195	Brazil	102
United Arab Emirates	183	South Korea	102
Brazil	146	Canada	101
Kuwait	144	Germany	96
Iran	137	Mexico	80

by war, natural disasters, or local economic unrest or economic downturns. Thus, countries like China, which has to import three times what it produces, are heavily reliant on oil-producing nations.

Natural Gas—The Clean Fuel? In recent decades there has been a major initiative to use more natural gas around the world because it is perceived as a much cleaner fuel than petroleum and coal. Indeed, when burned in power plants, natural gas releases around 50 times less carbon dioxide than coal. For many years, scientists and energy planners have developed a wide range of uses of natural gas because of its far cleaner emissions when compared with oil or coal.

Natural gas was not always such a popular energy source. For decades, natural gas was seen as a problem that oil producers needed to manage. It is found associated with petroleum and could not be as easily stored or moved as petroleum. Thus, it was often burned off as a byproduct. The same was true in other settings, such as in landfills or sewage treatment plants, where natural gas is a noxious byproduct of natural decomposition processes. It is only recently that we are finding ways to collect the vast amount of natural gas available from ancient reserves and from decomposing materials.

While it was not widely used in the past because it was seen as a nuisance, new questions about the use of natural gas come into play in our modern era due to its impact on our climate. As was noted in Chap. 5, methane is 25 times more powerful of a greenhouse gas as is carbon dioxide. Thus, one molecule of methane does the same amount of negative work to change our climate as one molecule of carbon dioxide. As a result, more and more climate scientists are raising alarm bells around the extraction, distribution, and use of natural gas.

Over the last few decades, hydraulic fracturing, or fracking, has been used to extract petroleum and natural gas from tight underground reserves. Extreme pressures exerted in the subsurface by fracking fluids crack the rock to release the energy sources. Of course, in such situations, some of the natural gas is released into the atmosphere where it can start to help to heat our atmosphere. The same thing also happens when natural gas or petroleum is extracted under normal extraction procedures involving traditional pumping. Fracking and regular oil extraction have an array of environmental challenges. However, this new challenge, the emission of natural gas into the atmosphere, is emerging as a critical issue for managing our climate challenge.

Once natural gas is extracted from the ground, it is moved around in a vast global network of pipelines. These pipelines may eventually lead directly to power plants, homes, or other locations where it is burned to create heat or electricity. This network is also supported by natural gas tanks which are transported by truck, rail, or ship. Anywhere along this supply network leaks can occur that do damage to the atmosphere. Recent research has demonstrated that there are over 600,000 leaks in the US pipeline system alone and that there are far more releases of natural gas through these systems than anyone recognized (Weller & others, 2020).

The US pipeline system is but one small part of the problem. There are hundreds of thousands of natural gas appliances of one form or another. When I did fieldwork in the Middle East, I regularly carried a small gas-powered cookstove to heat water for tea, and I now have a gas-powered clothes drier, stove, and home heating system. Each one of these has the potential for leaks. We have all smelled the telltale signs of natural gas during our day-to-day activities. There are natural gas leaks all around us and these leaks lead to changes in our atmosphere.

It is important to point out that natural gas is used more and more in vehicles. Many communities have converted fleet vehicles and buses to natural gas. Plus, many businesses that are highly transportation dependent, such as delivery companies, have extensive natural gas fleets. While the move to advance natural gas fleets across the world was done with the intent of seeking to reduce greenhouse gas emissions, many are now coming to the realization that the move has its own climate costs.

Communities that are particularly concerned about natural gas have started to ban the use of natural gas appliances and eliminate natural gas infrastructure in their communities. The world's top producers and consumers of natural gas are listed in Table 8.3. As can be seen when comparing Tables 8.2 and 8.3, there is a great deal of overlap overall between the oil-producing and natural gas-producing countries. This is logical because the reserves are often linked. Likewise, there is a similar overlap of oil-consuming and natural gas-consuming countries linked largely to population or the energy-intensive nature of the nations.

Table 8.3 List of top natural gas producers and natural gas consumers in the world as of 2019 (Enerdata, 2020b)

Top producers of natural gas	Amount (BCM)	Top consumers of natural gas	Amount (BCM)
United States	951	United States	877
Russia	740	Russia	501
Iran	240	China	304
Canada	183	Iran	226
China	175	Canada	129
Qatar	173	Japan	108
Australia	139	Saudi Arabia	98
Norway	118	Germany	95
Saudi Arabia	98	United Kingdom	80
Algeria	90	Mexico	77

The Pros and Cons of Nuclear Energy

Nuclear energy is a challenging topic because it is so multifaceted. On the one hand, nuclear energy has the potential to provide vast amount of energy with limited emissions. On the other hand, there are significant safety concerns over nuclear energy. Let's take a deeper look at both of these issues.

Nuclear Energy—The Emission Free Fuel? Nuclear energy typically refers to nuclear power plants that produce electricity for public consumption. Most nuclear power plants use the process of nuclear fission which involves splitting atoms to release energy. When the energy is released, heat is produced to create steam which turns a turbine to create electricity. Except for the heat source, nuclear power plants have similar processes as coal, oil, or natural gas power plants in that heat creates steam which turns a turbine which produces electricity.

In producing the heat via nuclear energy, nothing is burned. Thus, there are no releases of fossil fuels in the direct combustion of a heat source. Instead, particular radioactive elements are extracted from the earth and processed to create pellets that are inserted into fuel rods. The heat is produced as the radioactive elements in the pellets decay. Any greenhouse gas cost in this process is associated with the mining, production, and transport of these radioactive materials. As a result, nuclear energy is generally considered to be one of the main ways that we can maintain significant energy supplies while dramatically reducing greenhouse gas emissions.

Indeed, many climate policy analysts have argued that nuclear energy is the best way to address our current climate emergency (Biello, 2013). They make the point that other forms of green energy cannot ramp up fast enough to meet the demands of most societies and that the only way to rapidly cut emissions is to aggressively develop advanced nuclear power throughout the world. They also make the case that due to the public's fear over nuclear power that new, advanced, and safer nuclear technologies have not been developed.

This pro-nuclear approach does indeed hold promise. Most nuclear power plants around the world were built decades ago and utilize older technologies (Murray, 2019). Nuclear engineers have designed smaller, safer power stations, but they have not been built due to the expense and the overall public distaste for nuclear power plants near population centers. It is definitely possible to ramp up our use of nuclear energy to rapidly eliminate our use of fossil fuels.

Nuclear Energy—The Risks. The counter-argument to the pro-nuclear approach is the overall danger of nuclear power plants to the public. Two nuclear power plants have had meltdowns that permanently changed their regions—Chernobyl and Fukushima. The Chernobyl disaster occurred in 1986 in the former Soviet Union in what is now Ukraine. The accident, which was caused by human error, resulted in an explosion and fire which led to the release of huge amounts of radiation into the atmosphere over several days. Several people were killed initially and thousands died of unusual cancer illnesses in many downwind areas in Europe in subsequent years. Tens of thousands of people have been evacuated from the region and large areas of the land around the plant were initially closed off as cleanup ensued. Today, about 20 miles around the plant is set aside as an exclusion zone that has reverted to wilderness. The area is not expected to be safe for thousands of years.

The Fukushima nuclear disaster occurred as a result of a series of problems associated with a massive earthquake—with a magnitude of 9.0—that hit off the Japan coastline (Hutner, 2018). When the earthquake hit, the reactor shut down as part of normal operations under such circumstances. However, power to the plant subsequently shut down due to electrical grid disruptions. Under normal earthquake conditions, this would not be a problem. The reactor would continue to be cooled by water driven by pumps powered by generators on the plant. However, when a tsunami hit several minutes after the earthquake, the generators were engulfed with water and the reactors were no longer cooled. This unfortunately caused extreme heating resulting in three nuclear meltdowns and explosions. The earthquake and tsunami alone were major tragedies for Japan. Over 15,000 people died. Yet the long-term impacts of the nuclear meltdown are still felt due to ongoing challenges with trying to manage the damaged plant and the radiation in the region—including radiation released into surrounding coastal waters.

Besides the clear public health risks associated with nuclear power plant accidents, there are other concerns around issues of mining, nuclear waste, and nuclear weapons (Lynch & others, 2018). Perhaps the most pressing of these is the issue of nuclear waste associated with nuclear power plants. With time, the radioactive decay of the pellets used to heat the water in the plants declines to the point that significant heat is no longer generated. When this occurs, the pellets, which are stored in tubular rods, must be replaced. The spent rods, while no longer useful for nuclear energy, are still radioactive and hazardous to human health. They must be stored safely for thousands of years before they are no longer dangerous. While there are some reuse and recycling possibilities, the vast majority of nuclear power rods are stored at nuclear power plants or in long-term underground storage sites.

The challenge, of course, is how do we keep nuclear waste safe for thousands of years. Governments are not stable that long and there are always bad actors who could use the waste as a weapon against enemies. It is hard to imagine how we can effectively ensure that the waste will be kept from doing damage to future generations. In many ways, the nuclear waste issue is tied to the problems associated with nuclear bombs (Sasikumar, 2017). The processes associated with making fuel for nuclear power plants are not that different from those associated with making nuclear bombs. We have enough bombs to wipe out all life across the planet now. Advancing a nuclear power agenda to quickly get us away from fossil fuel use is attractive, but comes with significant risk.

Green Energy Use

Everyone loves the idea of green energy. We like the idea of a continuous free source of energy from nature that does not involve mining, pollution, or financial costs. We have been using green, or renewable, sources of energy for as long as we have been around on the planet. At the most basic level, wood is a renewable energy resource, if used wisely. But we all know that when it is overused, it leads to deforestation and environmental degradation. As with wood energy, there are distinct environmental costs associated with renewable energy use. While I think we would all agree that green energy is far better than dirty energy sources like coal and petroleum, it is inappropriate to assume that green energy has some limited environmental impacts.

This section focuses on three main forms of green energy: hydroelectric, wind, and solar. It is important to stress that not all areas of the world can utilize these green energy sources. For example, dreary cloudy areas might be more suitable for wind energy than solar energy and alpine regions might be more suitable for hydroelectric power generation than other forms of green energy. It is also important to stress that overall, energy use, both green and dirty, are increasing overall around the world. While we might think that green energy is replacing dirty forms of energy, the reality is that it is just allowing us collectively to utilize more energy. While many of us have embraced green energy while reducing our overall energy usage, collectively this is not the case across the planet. Therefore, as was stated at the outset of this section, we all like green energy—but the reality is we like energy period. We have not stopped expanding our planetary energy usage.

It is also important to note that hydroelectric generation is by far the greatest source of green energy. This makes sense when one considers that hydropower has been around since the late nineteenth century. The technology that drives large wind and solar power systems is relatively new. Nevertheless, the use of both of these technologies is increasing rapidly. Wind power production, however, produces nearly double the amount of electricity as solar energy (Table 8.4).

Table 8.4 Production of renewable energy in 2019. Note that hydroelectric energy is the largest source of renewable energy and that wind energy produces twice as much electricity as solar energy (Our World in Data, 2020b)

Green energy source	Production in 2019 (in TWh)
Hydroelectric energy	7027
Wind energy	1430
Solar energy	724
Total	9181

Hydroelectric Energy. Hydroelectricity is created when water, moving via the forces of gravity, moves through and turns a turbine to create energy. Large hydroelectric systems that can power whole cities require large differences in relief to create enough gravity force to turn turbines rapidly. As a result, most large hydroelectric systems are located in mountainous areas where rivers tumble down in waterfalls or rapids. In these locations, dams are built to create a reservoir which drains through the dam in large pipes which feed a turbine. There are thousands of these dams all over the world that are responsible for producing approximately 17% of the world's electricity. Thousands of dams are also being planned or built, mainly in Brazil, China, and East Asia. The largest producers of hydroelectric power are China, Brazil, the United States, and Canada. China is by far the largest producer in that it produces roughly three times more hydroelectric power than Brazil. While the United States is one of the largest producers of hydroelectric power, it is not planning many new dams due to a variety of issues including environmental concerns and community displacement. One of the reasons that China is such a large producer of hydroelectricity is that it is home to the largest hydroelectric dam—the Three Gorges Dam on the Yangtze River.

While hydroelectric energy is considered a green, renewable source of energy, it does have environmental impacts. Perhaps the most obvious of these is the flooding associated with the creation of the reservoir behind the dam. Vast areas are covered by water, thereby disrupting human settlement and natural ecosystems. The Three Gorges Dam floods over 1000 square kilometers of the earth's surface. The creation of the dam forced the relocation of over 1.2 million people. Another problem with dams is that they cause disruptions in the natural flow of rivers. Aquatic animals evolved with existing, natural stream flow. When dams are built, the natural life cycles of many of these animals are disrupted. While dams can control flooding under natural extreme conditions, they are susceptible to breaches under very unusual flooding conditions. We are already seeing a range of extreme weather due to global climate change and many communities downstream of dams are becoming more vulnerable to sudden dam failure. Finally, dams have a limited lifespan due to sediment build-up behind the dam. With time, sediment that would normally be carried through a river is trapped behind the dam walls. This lowers the amount of water that can be stored in the reservoir thereby lowering the effectiveness of the dam's ability to produce electricity. The Aswan High Dam, built on the Nile River in southern Egypt in the middle of the twentieth century, has unique sediment problems due to the large amount of sediment carried by the Nile. Today, over 70% of the reservoir has been filled with sediment (Moussa, 2013)

Wind Energy. Wind energy production has grown tremendously in recent years. Most of us have seen the vast fields of giant windmills that have sprung up all over the planet. The largest producers of wind energy are China, the United States, and Germany. However, there is an approximate 10% growth each year in the production of wind energy around the world driven in part by the leading producers of wind power. There is also considerable growth in Argentina, Chile, Mexico, Egypt, Iceland, Norway, Russia, Thailand, and Vietnam. In addition to expansive growth of wind farms on land, many new offshore wind farms are being built across the world. In the United States, extensive wind farms are being planned for the coastal waters off the Atlantic.

Windfarms do come at some environmental cost. Much has been written about their impact on birds, bats, and insects. While there is no doubt that there are fatalities associated with them, the impacts are not as large as originally thought. There is also concern over the environmental footprint of the windfarms. While the structures themselves have a relatively small footprint, they cannot be built over structures or in areas where they would do damage should there be structural failure. Thus, wind farms are typically found in rural areas where they can coexist with grazing animals or fields of crops. They are also found in areas with limited land use possibilities such as in arid lands or rugged terrain. They also have some noise associated with them. The deep whoosh of the blades can be irritating. There is also concern over the impact of offshore windfarms on marine ecosystems. In addition, the fishing industry is concerned about them for safety and also for how they impact natural fish stocks.

Table 8.5 Top solar energy-producing countries (Our World in Data, 2020b)

Country	Electrical production (GW)
China	205.5
United States	62.3
Japan	61.8
Germany	48.9
India	35.1

Solar Energy. Solar energy production has been increasing roughly at the same rate as wind energy (10% per year). China is by far the leader in solar energy production. However, the United States, Japan, Germany, and India are also major producers (Table 8.5). There are a number of countries that are rapidly developing solar energy capacity including Spain, Turkey, Egypt, the United Arab Emirates, Saudi Arabia, Vietnam, Australia, Brazil, and Mexico.

There are two basic types of solar energy used to generate electricity: photovoltaic and concentrated solar energy. In the photovoltaic process, electricity is produced within cells that convert photons to electric energy. What is particularly useful about photovoltaic cells is that they are highly portable and do not need to be placed within a power plant. Thus, the cells can be concentrated within solar panels that can be installed on buildings. These mini power plants allow any building to be its own electrical power plant. In recent years, there has been a big push to expand building rules to increase the production of solar electricity.

There are environmental impacts associated with this type of solar energy. While there are many types of solar cell technologies, many of them utilize rare earth elements or heavy metals to produce electricity. As a result, there are impacts associated with mining and waste management. Nevertheless, it is clear that the environmental impacts, particularly during a time of climate change, far outweigh the negatives.

Concentrated solar energy is not widely known because the power plants that utilize this technology are not that common. The way that they work is that a large array of mirrors concentrate the sun's heat at a single point. That heat is used to create steam, which then turns a turbine that produces electricity. Spain and the United States are the largest producers of concentrated solar energy, but this technology is diffusing rapidly. In 1990 there were 300 concentrated solar energy plants. Today, there are over 6000 of them throughout the world. The challenge with this type of solar energy is that it requires a relatively large open area to build up the mirror arrays needed to direct heat. In addition, this type of system is most suitable for areas with limited cloud cover. As a result, most of these systems are located in arid or semiarid environments in the lower latitudes.

Going off the Grid

Given the range of emerging technologies, more and more people are trying to get off the grid. Getting off the grid means getting away from dependence on power supplied by power companies. Homeowners can produce their own electricity using solar cells or small windmills. Of course, the challenge is that this type of energy is not always available. As a result, getting off the grid is often a term meant to indicate partial independence from the regional electric power grid. There are emerging battery technologies that would allow homeowners to store excess energy for use when solar or wind is unavailable. However, these battery systems are expensive and not widely available.

Some communities have policies that require power companies to pay for excess energy produced by homeowners' green energy systems. Thus, on sunny days when solar panels are churning out electricity, power companies buy the excess electricity for redistribution on the grid. Power companies have tried to limit these policies in recent years because of the loss of revenue.

How You Can Make a Difference

Everyone loves the idea of green energy, but it takes a personal commitment to make a difference in our own lives. Table 8.6 outlines the simple, hard/expensive, and innovative/life-changing ways you can make a difference by advancing a green energy agenda. The easiest thing you

Table 8.6 Simple, hard/expensive, and innovative/life-changing approaches to infusing green energy in your life

	Ways to embrace green energy
Simple	Purchase green energy credits from your power company
Hard/expensive	Install solar panels or a small windmill on your home
Innovative/life-changing	Commit to going off the grid

can do is to purchase green energy credits. There are a number of ways to do this. Most power companies allow you to pay a little extra on your bill which allows them to invest in green energy. This may cost a few dollars more a month, but it will help promote green energy in your region. You can also buy carbon credits. Chapter 5 discussed carbon credits in detail. The way that this works is that you calculate the amount of dirty energy you utilize and pay a company to counter your carbon production with green projects that either sequester carbon or advance green energy projects.

A harder and more expensive approach is to install some type of green energy producing appliance on your home. You actually have lots of options. Many companies around the world sell solar panels that can supplement your electrical usage. Also, there are many small windmills that can be easily installed in some areas. This option is not available for everyone. Some areas have zoning regulations that prevent the installation of solar panels or windmills and many people rent their homes and cannot make changes to their infrastructure. Nevertheless, installation of home-based green energy is increasing worldwide and many of us have the ability to make the transition.

To fully embrace green energy, one needs to commit to going off the grid. This involves committing to producing the amount of energy you would normally consume through renewable energy sources—typically wind or solar energy. You can cut yourself entirely off the grid by storing excess energy in batteries for use when wind or solar is not available. If you stay on the grid, you can push your excess energy onto the grid during windy or sunny days and pull from the grid at night or during calm periods without needing batteries.

Committing to Green Energy with Overconsumption in Mind

As stated earlier in this chapter, the worldwide growth in green energy has not led to a worldwide decline in dirty energy production. As a result, it is important to keep in mind that throughout the world we need to find ways to cut energy consumption and improve energy efficiency. We need to consume less energy overall in order to limit the use of dirty energy sources. While the move to green energy is a step in the right direction, it is not a cure to our energy or climate change problems. We all need to look holistically at how we can move forward to aggressively ramp up our production of green renewable energy while cutting our use of dirty energy around the world.

References

Biello, D. (2013, December 12). How nuclear power can stop global warming. *Scientific American*. https://www.scientificamerican.com/article/how-nuclear-power-can-stop-global-warming/

Enerdata. (2020a). Coal and lignite domestic consumption. Retrieved December 5, 2020, from https://yearbook.enerdata.net/coal-lignite/coal-world-consumption-data.html

Enerdata. (2020b). Natural gas domestic consumption. Retrieved December 5, 2020, from https://yearbook.enerdata.net/natural-gas/gas-consumption-data.html

Hutner, H. (2018). Japanese women and antinuclear activism after the Fukushima accident. In R. Brinkmann & S. J. Garren (Eds.), *The Palgrave handbook of sustainability: Case studies and practical solutions* (pp. 283–298). Palgrave Macmillan.

Lynch, S. N., Lambert, L., & Bing, C. (2018, October 4). U.S. indicts Russians in hacking of nuclear company Westinghouse. *Reuters*. Retrieved June 1, 2019, from https://www.reuters.com/article/us-usa-russia-cyber/u-s-indicts-seven-russians-for-hacking-nuclear-company-westinghouse-idUSKCN1ME1U6?il=0

Moussa, A. M. A. (2013). Predicting the deposition in the Aswan High Dam Reservoir using a 2-D model. *Ain Shams Engineering Journal, 4*, 143–153.

Murray, L. (2019). The need to rethink German nuclear power. *The Electricity Journal, 32*, 13–19.

Our World in Data. (2020a). Energy. Retrieved December 5, 2020, from https://ourworldindata.org/energy

Our World in Data. (2020b). Renewable energy. Retrieved December 5, 2020, from https://ourworldindata.org/renewable-energy

Sasikumar, K. (2017). After nuclear midnight: The impact of a nuclear war on India and Pakistan. *Bulletin of the Atomic Scientists, 73*, 226–232.

Weller, Z. D., Hamburg, S. P., & von Fischer, J. C. (2020). A national estimate of methane leakage from pipeline mains in natural gas local distribution systems. *Environmental Science and Technology, 54*, 8958–8967.

9

Protecting Our Water Resources

Introduction

There is an old saying that water is life. If that is the case, why do we treat our natural water systems so badly? We certainly have water all around us and it is hard to imagine that the supply has limits. There are vast areas of oceans, lakes, rivers, and wetlands. Plus, we have extensive underground reserves of water in aquifers. Yet throughout the world, there are serious water shortages and drinking water is degraded to the point that it is unusable. We are destroying the very systems that support us. This chapter will look at issues of both water quantity and water quality to better understand what we can do to try to improve the sustainability of water on our planet.

There is evidence all around us for the problems associated with water. Perhaps the most shocking piece of evidence is the destruction of the Aral Sea which was located between Kazakhstan and Uzbekistan in southwest Asia. Agricultural development schemes in the second half of the twentieth century diverted water away from the lake to agricultural fields. As a result, the water level dropped steadily and today only a few small remnant lakes remain. The Aral Sea essentially disappeared in just a few decades. This is just one example of how our modern consumerist culture is destroying water resources around the world. In every corner of our planet, water is under threat in some way. We need to find ways to protect our precious water systems in order to ensure that they can be around for generations to come.

Understanding the Water Cycle and How We Modify It

The water cycle is one of the most fundamental natural planetary cycles, along with the carbon cycle and the rock cycle. The cycle shows (Fig. 9.1) that water can move across various phases through natural surface processes. Water can evaporate from surface water to enter the atmosphere to produce atmospheric moisture and clouds. The clouds can produce some

Fig. 9.1 Niagara Falls shows the water cycle in all of its glory

form of precipitation such as rain or snow. That moisture can then run off across land to produce surface water or run into the ground to produce groundwater. From there, the cycle starts over again. It is important to note that while the water cycle is extremely fast moving as is evidenced by the movement of severe weather systems, vast amounts of water are stored in oceans, glaciers, groundwater, and surface water.

We interfere with the water cycle in many ways. Perhaps the most fundamental way we are modifying the cycle right now is via global climate change. Our release of huge amounts of greenhouse gases since the industrial revolution has changed our atmospheric chemistry and warmed the atmosphere. What this means is that our atmosphere can hold more moisture which means that we have more severe storms, extreme rainfall events, and changing weather patterns. We also have rising sea levels caused in part by melting glaciers but also by the increased oceanic volume driven by warming seas. When a unit volume of liquid warms, it increases in size. Thus, even without adding any water to the oceans from melting glaciers, we would expect to see sea levels rise as ocean water temperatures increase.

We also interfere with the water cycle by moving water around from one location to another. The Aral Sea example in the introduction is but one way we change the natural distribution of water. We do this through big schemes, like the redistribution of Colorado River water to thirsty western cities in the United States, and through small schemes like the draining of a wetland to build a vacation home. We also withdraw huge amounts of water for municipal drinking water, agriculture, and industry. There are few natural waterways or coastlines on the planet that have not been modified in some way.

We also change the chemistry of the water through widespread water pollution—both point pollution, or pollution that is released at a single source, and non-point pollution, or pollution that is regionally released such as with agricultural nutrient pollution. All over the world water chemistry is altered in most water ways in some ways. Our ability to transform our planet in big and small ways has led to profound changes in water chemistry that directly impacts ecosystems and human survival.

Managing Water Quantity

We are all responsible in some way for local, regional, and global water problems. In terms of actual water consumption, we actually drink relatively little water. In the United States, we use 45% of the water we withdraw for power plant cooling (45%). Irrigation accounts for the next highest use (32%). We only use 12% for our domestic supply. We use most of this domestic water to flush toilets, take showers, in faucets, and clothes washing. Roughly 17% of our domestic water is lost to leakage (EPA, 2020). We also are responsible for the water associated with the food we eat and the materials we consume. If we look at national annual water consumption (Table 9.1), Turkmenistan is the largest consumer of the world of water resources on a per capita basis. This is largely because the country is arid and it relies heavily on irrigation for agricultural production. Indeed, most of the countries on the list have arid or semiarid areas which require considerable amounts of water for irrigation. The numbers presented in Table 9.1 are heavily impacted by agricultural, industrial, and mining water uses.

Household Water Uses. What was not evident in Table 9.1 is that there are big differences in the amount of water consumed by households around the world. People in some nations are big per capita users of household water while people in other nations are not. In the United States, where green lawns, regularly cleaned clothes, and daily or twice

Table 9.1 The list of the countries with the highest annual per capita water withdrawal (Our World in Data, 2020)

Country	Annual per capita water withdrawal in m^3/person
Turkmenistan	5753
Iraq	2646
Chile	2152
Uzbekistan	2473
Guyana	1905
Tajikistan	1619
United States	1543
Kyrgyzstan	1531
Estonia	1310
Iran	1301

daily showers are the norm, household water consumption is quite high. The consumption is driven, in part, by well-functioning water distribution systems that bring cheap sources of water to homes. In contrast, some countries do not have regular water distribution systems and water is delivered by truck to rooftop or underground tanks. In these types of places, where water is expensive and scarce, household water consumption is very low compared to the United States.

But even in the United States, water consumption can vary greatly. If you live in an apartment or condominium, your water consumption is likely relatively low. Contrast this water use to a large multi-acre estate with extensive lawns. In wealthy enclaves of southern California, Saudi Arabia, and Long Island, New York, some property owners feel that they should be able to use as much water as they need for lawns or other purposes as long as they can pay the water bill. Of course, the idea that one person should have disproportionate access to a single resource is not ecologically sound and many water agencies have limits on how much water a single property owner can use.

Residential water policies vary from place to place around the planet. In some places, water is seen as a basic human right and is provided at a reasonable price by local governments. There are two broad approaches to pricing water. One approach charges customers the same amount per volume of water regardless of how much is used. In this model, light users pay the same price per gallon as heavy users. Another approach uses a conservation price model which charges more per gallon as use goes up. Thus, if you are a heavy user, you are charged more to use lots of water to water your lawn or feed your pool.

This conservation approach goes against the traditional economic approaches that tend to charge less for purchasing lots of a particular product. Because water infrastructure can be expensive for communities to maintain, there has been a major move as of late to advance privatized water systems in many locations. What this means is that for-profit companies take over the management of local water systems and make money off of selling water to local customers. While privatized systems usually have some degree of conservation pricing associated with water delivery, at the end of the day, these companies are in the business of selling water.

There is considerable pushback against privatization of water. If we all agree that water is a basic human right, why would communities give over their water resource management to a private company? What happens to people who cannot afford access to local water supplies? In addition, water resources are commons resources. Overuse or bad water management will lead to deteriorating water conditions for everyone. Do we want to turn over the management of commons resources to for-profit companies?

While local governments are typically in charge of water supplies, or have contracted with private companies to provide a steady supply of water to residents, we do have, as individuals and as households, some degree of agency in how we utilize the water provided to us. New technology such as low-flow toilets, innovative dishwashers and clothes washers, and conservation shower heads can significantly reduce household water usage. In addition, we can limit the need for outdoor watering by planting landscaping that is in tune with our local ecology so that watering is only needed during extreme dry periods. In addition, it is important to maintain household water systems to avoid leaks.

Agricultural Uses. Our modern system of agriculture is highly out of whack with natural ecosystems. We have the technology to transform deserts into farm fields. Many regions of the world are in the midst of using scarce water resources for agricultural transformations as part of local or regional development schemes. In some places of the world, such as in Southwest Asia and the Aral Sea, this leads to long-term disruptions in natural surface water systems. In other cases, water is mined from aquifers that contain water that has been in place for thousands of years. The noted Ogallala Aquifer, located under sections of the central portion of the United States, is nearly out of water. Farmers over the last century have pumped this aquifer system to grow wheat and other crops across a vast semiarid landscape. These agricultural transformations come at distinct environmental costs.

Agricultural processing is also responsible for a great deal of water use. Most of you reading this book have access to produce that is washed prior to leaving the farm field. It is often washed again prior to packing. Washing is important for hygienic purposes, but our modern expectation of what is clean is very different from what our parents or grandparents expected. It is not uncommon to buy a bag of lettuce that is "triple washed" or to have vegetables misted within a produce section of a grocery store to keep them moist and fresh looking. These supermarket vegetables are often very different from what we would find at a farmers' market or from what we would grow in our gardens. We are rather forgiving of the cleanliness or look of these vegetables and have far higher expectations when we purchase them in a grocery store. Vegetables that do not have the look that consumers desire are pulled aside during processing at farms or warehouses which leads to food waste and ultimately a waste of water.

It is important to highlight that our worldwide shipping of fresh agricultural products is really an exercise in shipping water. Most fruits and vegetables are high in water content and thus when we ship produce from, say, Peru to China, we are transferring water, in the form of produce, from one location to another. Because water is heavy, there are significant carbon costs associated with the international fresh produce trade. This carbon cost is one of the reasons why there is so much interest in creating local food systems.

Mining. Most mining activities are very water intensive. At the most basic level, the mining process involves bringing mineral ores to the surface of the earth in raw form where they are concentrated and processed for creating useful items. The process of concentrating and processing the ores is what uses the most water. There are some types of mining that are particularly problematic for water use and for associated ecosystems that rely on water. I want to focus on just two of them: phosphate mining in Florida and tar sand mining in Alberta.

The phosphate mining industry in Florida is focused on extracting material from a particular geologic layer called the Bone Valley Formation (FIPR Institute, 2020). What makes this Formation so productive for those seeking to mine phosphate is that it contains great amounts of Cenozoic Era animal bones which contain the mineral apatite which is made up of calcium phosphate. The apatite, once mined, is processed to produce phosphorus-rich fertilizers that are used on farms, gardens, and lawns. The challenge with mining the Bone Valley Formation is that there is another layer of sediment on top of it that must be removed and stored to get at the phosphate ore. Plus, the phosphate ore exists as a relatively thin flat layer that must be strip mined. The strip mining causes significant disruption of the landscape of west-central Florida, one of the most productive phosphate zones in the world. However, huge amounts of water are consumed in the process because the overburden is transported away using hydraulic slurries and the phosphate-rich material is transported to processing plants using hydraulic slurries in pipes. Huge amount of sediment-rich wastewater is produced in the process. Plus, the processing of phosphate produces significant water quality problems.

The tar sands in Alberta are mined in similar fashion using strip-mining techniques. However, to get the tar out of the sand, the sand is heated in the presence of water which helps to loosen the tar so it can be separated. The end result is sediment, contaminated water, and petroleum. The water systems of Alberta have been significantly disrupted through the mining process.

Running Out of Water

Some communities around the world are running out of water. Sana'a, the capital of Yemen, is very rapidly approaching a crisis because it is using far more water than the local sources can provide. As a poor, arid nation, there are few options and it is likely that there will be widespread problems and even out-migration as a result of this challenge. Many areas of central Asia and parts of Australia, Africa, and South America are also having serious water challenges.

Some wealthy communities have sought to fix the problem by advancing new technologies to try to create new water from the sea. A range of new desalination plants have emerged in a number of locations to try to provide freshwater to thirsty populations. The largest such plant in the United States provides drinking water to the Tampa Bay region which had devastated local aquifers through overuse to the point that saltwater intrusion made many of the aquifers unusable for generations (Tampa Bay Water, 2020). Other regions have used a range of water harvesting or water conservation technologies to try to limit water usage. Many areas with water shortages, such as California, Spain, and Israel, have developed extensive drip irrigation technologies for agricultural fields.

Water Quality

One of the growing concerns associated with water is that there are many water crises that are confronting the world's populations. When looking at the global water budget overall, the vast majority of water, around 96.5% of it, is in oceans. The remainder of it is tied up in glaciers, atmospheric vapor, rivers, lakes, and aquifers. Only about 1.5% of all of the Earth's water can actually be used for useful purposes such as drinking water and irrigation—and this water is incredibly vulnerable to contamination due to our actions.

Think about the freshwater sources where you live. We all imbibe drinking water that comes from groundwater or surface water sources. Some of us even get our drinking water from the ocean from a desalination process. I am writing this chapter in DeKalb, Illinois, where we get the vast majority of our water from groundwater sources. There are multiple aquifers in the region, but two that are productive for drinking water. Because it is a highly agricultural region, the water is regularly tested for pesticides, herbicides, and a variety of nutrients. The water quality meets or exceeds all federal and state drinking water standards. However, all aquifers are vulnerable to contamination from spills. We are very lucky in my region to have limited groundwater problems. Of course, this is not always the case. Some communities rely on a range of aquifers or surface waters that are problematic and that require extensive treatment to make them potable. In some parts of the world, the water supply is iffy and the public needs to do some degree of home treatment such as boiling or using bleach additives in order to make the water safe for drinking.

There are four main types of drinking water contaminants: physical, chemical, biological, and radiological. The physical contaminants are those contaminants that somehow impact the color or appearance of water. Many communities around the world draw their drinking water from surface water that is colored with organic tannic acids that stain the water a brownish red. These chemicals can also create an unpleasant taste to the water and many water utilities treat the water to remove the color and taste. Some places have cloudy water due to dissolved sediment in the water.

Chemical contaminants are highly problematic in our world's water supply. They vary widely and include a range of organic and inorganic chemicals like nitrogen, lead, and solvents. There are multiple sources of these chemicals. For example, point sources, such as a leaking underground storage tank of solvents and an oil spill of a river could widely contaminate groundwater and surface water systems. In contrast, non-point sources, such as agricultural fertilizer or arsenic-containing emissions from power plants, can widely contaminate broad areas.

What is particularly challenging about water contaminants is that we do not see them and do not know if they are in our drinking water. In addition, low levels over time can have devastating impacts. Plus, while municipalities may be delivering clean water to the underground infrastructure, contaminants can enter from older or decaying infrastructure as was seen in Flint, Michigan. There, clean, but inappropriately treated water, released metals from old pipes causing lead poisoning. The issue of chemical contamination of water is complex and there are new emerging chemicals, such as pharmaceuticals and nanoparticles, that continue to cause challenges around the world.

Biological contaminants are those that cause illnesses such as Cryptosporidium and E. coli. If water is not properly treated a range of health issues can occur. Indeed, over three million people a year die from some sort of water-borne illness. Radiological contaminants include natural contaminants such as radon as well as unnatural contaminants associated with nuclear releases. The most recent Fukushima nuclear disaster is responsible for the contamination of millions of gallons of water that were released on the land surface and into the ocean and there is great concern about the long-term impacts of this contamination.

Managing Water Quality

Over the years, governments have developed ways to manage water quality problems by setting limits on emissions of particular chemicals or outright bans of particular pollutants. The Clean Water Act in the United States allowed the Environmental Protection Agency (EPA) to set a variety of standards for water quality for a variety of uses. The EPA sets standards for drinking water and recreational water for example. The standards provide threshold limits for a range of organic and inorganic chemicals. When contaminants exceed these threshold limits, the EPA takes action. The water is regulated when it is in the environment and there are also regulations about what can be released into the environment.

While setting standards is a great thing to do, there are a number of challenges to managing water quality listed below.

1. Water testing is expensive. Basic water testing of a few attributes such as nitrate and pH may cost $10 or $20. However, if you are testing for a range of pollutants, a single water test can be hundreds of dollars. Thus, if you are doing any type of regional sampling, testing can cost thousands of dollars.
2. Pollutants are invisible. Because most pollutants are not visible in water, it is difficult to know the range of pollutants that could be in the water. As a result, water testing could miss some contaminants that could be problematic.
3. There are many emerging pollutants that are not part of standard testing. We now regularly produce new chemicals that make their way to the marketplace without significant environmental testing. More about this issue in the next section.
4. Many parts of the world have specific regulations, but there is lax enforcement. It is expensive to maintain a regulatory body and to test water on a regular basis. As a result, many parts of the world have limited governmental oversight of water pollution.

It is relatively easy to ban pollutants or limit their release at a particular site. However, it is much harder to manage non-point pollutants such as fertilizers. All over the world, water bodies have become polluted with nutrients that run off of farm fields, golf courses, and lawns. Perhaps the best evidence of this are the numerous "dead zones" that have formed in coastal waters over the last century off the coast of agricultural regions. One of the most significant of these forms is in the Gulf of Mexico near the Mississippi River delta. Large amounts of nitrogen and phosphorus enter the Gulf causing algal blooms. These blooms subsequently cause eutrophication or deoxygenation of the water. As a result, fish in the region die due to the lack of oxygen.

The non-point nutrient problem is not just an issue for coastal waters. All over the world we are seeing slow and steady rises in nutrient levels in surface water and groundwater. One of the largest concentrations of freshwater springs in the world is located in Florida. Here, tens of thousands of gallons of water coming from deep underground emerge into pools that form lakes or streams. This water was once considered the cleanest in the world. Today, many of the springs are contaminated with nutrients. One of them, Sulphur Springs, in Tampa, is so contaminated that the spring is no longer open to the public (Brinkmann, 2013). At one time, Sulphur Springs was an international tourist attraction. Today, it is fenced so that people are not exposed to the polluted water.

Stormwater is another source of water pollution. During rain events, water picks up lots of sediment and pollutants and it flows toward streams. Typically, there is a "first flush" of pollutants that enters a stream during the first part of a rainstorm. In cities, this first flush carries all of the debris that can settle between rain storms—pet wastes, ash, atmospheric dust, litter, and so on. In rural areas, this first flush often contains very high levels of nutrients running off of farm fields. In either case, stormwater is a serious environmental problem. In addition, stormwater can cause flooding.

Some communities try to manage stormwater by treating it like sewage. They divert stormwater via sewers to sewage treatment plants. Other communities have built diversion systems that take stormwater to water retention ponds where contaminants can be filtered by soils or sediment and where wetland plants can retain the nutrients. However, in most places, stormwater enters surface water bodies untreated.

Another important part of managing water quality is through sewage treatment. Sewage can be a major health problem. In the mid-1800s in England, the father of public health, John Snow, traced cholera outbreaks to drinking water systems near privies (Brinkmann and Tobin, 2001). As a result, there have been significant efforts put forward to improve sewage management within comprehensive wastewater systems. While rural communities in many parts of the world are still vulnerable to health problems due to reliance on privies or septic systems, most major cities have developed scientific approaches to wastewater management via sewage treatment plants.

Sewage treatment plants process wastewater by removing solids and by using natural bacteria to reduce the nutrient content. Once treated, the water is released into the environment and the solids are landfilled or dried and used for other applications such as fertilizer. The City of Milwaukee diverts some of their sewage into a commercial product called Milorganite which is a nitrogen-rich fertilizer that is often used on lawns and golf courses.

Emerging Water Pollutants

Even though many countries have some degree of water quality standards and management, they cannot keep up with the range of emerging pollutants that can cause problems. Some of the classes of emerging pollutants are listed below.

1. *Microplastics*. Plastic is showing up in drinking water all over the world. Some of the plastic derives from microbeads which are used as abrasives in a variety of products such as soap and lotions. They end up in wastewater streams and are typically not captured in any wastewater processing and are released into the environment where they can be picked up in our drinking water. In addition, plastic can naturally break down into small pieces in nature. For most cases, it can take decades to chemically break down so it remains persistent in the environment.
2. *Pharmaceuticals*. We have seen a massive increase in the range and volume of pharmaceuticals used around the world by humans and animals. The drugs we take make their way through our digestive systems to be released into wastewater. Our wastewater systems are not effective at removing the drugs. As a result, we can regularly find things like anti-depression drugs in surface waters. We also give lots of pharmaceuticals to farm animals. For example, we find lots of hormones in waters near dairy farms. We are even finding increased levels

of caffeine in water due to our cultures' growing penchant for coffee. Because of the pollution associated with the release of pharmaceuticals, communities have developed drop off sites for old or unused pharmaceuticals so that they are not flushed down toilets.
3. *Personal Care Products.* One of the most vexing classes of water pollutants comes from our use of personal care products such as cosmetics, lotions, sunscreens, and soaps. We would not normally eat or drink these products, but the chemicals that make up these products are showing up more and more in our drinking water and we do not yet know what the long-term impact will be. We do know that sunscreen that washes off of swimmers has caused permanent DNA damage to corals and many areas now ban the use of sunscreen when swimming.
4. *Nanoparticles.* We regularly produce and release nanoparticles in a variety of ways. For example, lots of nanoparticles are produced in spray painting or coating. Nanoparticles are also released in burning fossil fuels or in candles. While we regularly think of these materials as air pollutants, they can also enter water systems. Their impact is largely unknown.

Ways to Conserve and Protect Water

There are many things that we can do as individuals, families, and organizations to try to conserve water and protect water from pollution. Table 9.2 lists some of the ways that you can take action. The simplest way to take action is to take control of your water use in your home and make changes to how you consume water through your day-to-day activities. For example, putting a brick in a toilet tank saves that brick's volume of water with every flush. You can also install inexpensive lower-flow shower heads and cut shower time. In the kitchen you can prepare fruits and vegetables without letting the water run and you can also shut the faucet while brushing your teeth in the bathroom. Outside, you can water your lawn and plants only when needed. To take it one step further, you can install water-conserving appliances (dishwasher and washing machine) and replace old toilets with water-conserving toilets. Outdoors,

Table 9.2 Simple, hard/expensive, and innovative/life-changing approaches to conserve water and reduce water pollution

	Ways to conserve water	Ways to reduce water pollution
Simple	Do the basics of water conservation: modify toilet tank, limit amount of water used in bathing by cutting shower time and by adding low-flow shower heads, fix leaks, do not let water run when cleaning vegetables or brushing teeth, water plants only when needed.	Do not purchase cleaning products that are harmful to the environment. Do not release any liquid wastes (oils, gasoline, etc.) into the environment. Dispose of prescription drugs appropriately.
Hard/ expensive	Install water-conserving appliances and toilets (e.g., dishwashers and clothes washers). Limit lawn areas for watering.	Eliminate the use of outdoor chemical fertilizers, herbicides, and pesticides.
Innovative/ life-changing	Install xeriscape landscaping using local vegetation. Install permeable pavement. Install rain barrels for capturing water that can be used for watering plants, washing clothes, or flushing toilets.	Get to know your watershed and the major polluters. Work to promote policies that will eliminate or reduce emissions of pollution into surface or ground water.

you can also limit the areas of the lawn or garden that will receive regular watering. To deepen your commitment to water conservation, you can replace your lawn with local vegetation so that it needs little to no watering since it will be in tune with the local ecology. You can also replace impermeable surfaces, such as driveways, with permeable pavement. You can also develop a water collection system for the home that captures rainwater that would normally run off your home during rainstorms for use in a greywater system for flushing toilets, washing clothes, or watering plants.

To do your part to avoid water pollution, it is important to purchase and use green cleaning products and soaps and to avoid dumping liquid waste such as oils, gasoline, and grease into the environment. In addition, over-the-counter and prescription drugs should be disposed of appropriately and not flushed down the toilet. To deepen your commitment to

Fig. 9.2 The dry uplands near San Antonio go through boom and bust water cycles. As a result, the Edwards Aquifer Authority helps to manage water within the Edwards Aquifer watershed

cutting water pollution, cut your use of outdoor fertilizers, herbicides, and pesticides. Finally, one of the hardest things to do is to get to know your watershed and to understand how it is threatened by polluters. Watersheds are areas of land that are drained by a particular stream system. We all live within a watershed. For example, I currently live within the Kishwaukee River Watershed. Any pollutant that is released in the watershed has the potential to enter the Kishwaukee River or the groundwater systems in the region. As a result, it is common for watersheds to have watershed management organizations or community groups that work to protect them. Find your watershed and learn how you can get involved in reducing water pollution in your region (Fig. 9.2).

References

Brinkmann, R. (2013). *Florida sinkholes: Science and policy*. University Press of Florida.
Brinkmann, R., & Tobin, G. A. (2001). *Urban sediment pollution*. Springer.
EPA. (2020). How we use water. Retrieved December 5, 2020, from https://www.epa.gov/watersense/how-we-use-water
FIPR Institute. (2020). Mining and beneficiation. Retrieved December 5, 2020, from https://fipr.floridapoly.edu/research/mining-and-beneficiation-mineral-processing.php
Our World in Data. (2020). Water use and stress. Retrieved December 5, 2020, from https://ourworldindata.org/water-use-stress
Tampa Bay Water. (2020). Tampa Bay seawater desalination. Retrieved December 1, 2020, from https://www.tampabaywater.org/tampa-bay-seawater-desalination

10

Dealing with the Garbage Around Us

Introduction

Back in the 1960s, the United States had a litter problem. The Interstate Highway System was relatively new and growing, but the highways were full of the society's detritus. Cans, bottles, and a variety of garbage were thrown from cars to the side of the road where it created not only visual blight, but also serious environmental challenges from the number of pollutants found in the roadside trash. At the same time, the wide availability of consumer goods was creating a new waste problem. We were throwing away much more stuff and existing dumps were ill equipped to deal with it all. Plus, as we developed new chemicals, they started to get intermingled with our waste and it became clear that we needed to develop more scientific ways to deal with garbage.

This era saw the development of new rules and regulations to deal with these waste problems. For example, the Highway Beautification Act was passed by the US Congress in 1965 to address a variety of pollution issues associated with US roadways and in 1976 the United States passed the Resource Conservation and Recovery Act that started to regulate how we manage solid waste. All over the world, new rules around waste, litter, and pollution were developed to try to deal with the new realities of garbage.

This was a profound change in how we managed waste. No longer was our waste an individual or household problem. Now, it was something that was managed by local, state, or federal governments. Prior to this, households regularly burned garbage or buried it in garbage pits. Or, they composted organic materials and recycled or reused inorganic waste. Some took their garbage to a local dump where it might be burned or in some other way dealt with so it was not a major physical burden on the landscape.

By making garbage pickup so easy in today's culture, we have essentially relegated the responsibility of our waste to others. In our current culture, it does not matter how much you throw away. Someone else will deal with it. As a result, we have created a massive infrastructure around waste throughout the world and the amount of waste we create is increasing.

Types of Waste

There are dozens of types of waste. This section will review hazardous waste, electronic waste, industrial waste, medical waste, liquid waste, agricultural waste, and household waste. This is not a full list of waste types, but they are among the most common.

Hazardous Waste. This type of waste is somehow harmful for the environment or for human exposure. There are a range of sources of hazardous waste in our culture. Most of us have some type of product in our home that could become a type of waste called household hazardous waste. A list of common household hazardous waste sources is found in Table 10.1. While it is difficult to regulate a household, companies and organizations that use hazardous wastes regularly must dispose of hazardous waste appropriately. The recent disaster in Beirut, Lebanon, showed the devastating impacts of improper storage of waste. In 2020, ammonium nitrate, which was essentially an improperly stored waste product, exploded to kill and injure hundreds and to cause billions of dollars in property damage in Lebanon's leading city. The destruction adjacent to

Table 10.1 List of common household hazardous wastes

Waste associated with cars	Oil
	Gasoline
	Antifreeze
	Car cleaning products
Waste associated with interior design	Paint
	Varnish
	Coatings
	Glues
	Caulks
Cleaning products	Drain cleaner
	Toilet and tile cleaner
	Rust remover
	Carpet cleaner
Outdoor chemicals	Fertilizers
	Pesticides
	Herbicides
Other materials	Batteries
	Fluorescent light bulbs
	Pool cleaners
	Pharmaceuticals and over-the-counter drugs

this ancient port showed how explosive waste can devastate an area. However, most hazardous waste is regulated so that it does not enter ecosystems or our drinking water.

Electronic Waste. We are deep into a technological era; however, there is a significant digital divide across the planet. North America, Australia, and Europe have the most access to technology and the amount of electronic waste produced in these places is vast. However, it may be a surprise that these places do not produce the most electronic waste. China, with its large population, is the e-waste leader with the United States, Japan, and India also in the top four producers. Electronic waste comes in many forms. It can be old phones, old sound systems, defunct computers, and any other type of electronics.

The problem with electronic waste is that electronic components contain a range of hazardous chemicals that can range from rare earth elements to heavy metals. Some of the components are valuable which makes electronics recycling a potentially lucrative business. However, it is difficult to recycle all parts of electronic equipment because not all components are that valuable. For example, a desktop computer contains a large amount of low-value plastic and glass. It is not uncommon for some pieces of electronics to get diverted for reuse or recycling while other parts enter the waste stream. The challenge with e-waste is that, overall, recycling of the equipment does not return much money. As a result, large amounts of it get sent to poor countries with low labor costs where it can be mishandled or where it can be diverted into local waste streams. This dumping of electronic waste has caused problems in many countries.

Industrial Waste. Industrial waste includes a large number of different types of waste that are produced in industrial processes. What is key is that industries tend to produce the same type of waste over time so that it is relatively easy to develop sound waste strategies. For example, a wood furniture manufacturer may produce a large amount of wood scraps that could be diverted for recycling or reuse. Thus, with sound planning and intervention, industrial waste, due to its temporal homogeneity, is a good candidate for diversion from the waste stream.

Medical Waste. Medical waste includes a range of material that is in some way dangerous or contaminated with biohazards. It may include things like used sharp syringes or cutting implements and biomaterials like blood or tissue samples and lab test residue. The challenge with some forms of medical waste is that it can be harmful to humans upon exposure due to the fact that it may contain pathogens, mutagens, or carcinogens. As a result, medical waste must be handled separate from other forms of waste. Typically, medical waste is burned at an off-site facility or is autoclaved, or treated with heat and steam for disinfection, at an onsite facility. In most developed nations, there is a highly regulated pick up and disposal bureaucracy in place to manage the waste. Even small medical facilities like doctor's offices have regular pick up of medical waste. However, medical waste can be a vexing waste problem in developing countries without the appropriate handling sites and waste treatment systems. A new emerging form of medical waste around the world is the disposable face mask. With COVID-19, they have become common forms of litter.

Liquid Waste. There are a range of liquid wastes that we manage. They include materials like sewage, animal farm waste, fats and greases, and liquid slurries from manufacturing or mining. This book covered sewage and liquid mining wastes in the last chapter. The remainder, however, deserve a bit more focus. Animal farm wastes, such as animal manures, are often collected for treatment using similar processes as is done in sewage treatment plants. These treatment facilities are found at large farms. Some places even require some degree of treatment for relatively small holdings to avoid pollution of waterways. Fats and greases, which are often collected from restaurants, often require special handling because they can clog or otherwise damage sewage systems. In recent years, there has been great interest in recycling waste oil and grease as a fuel.

Agricultural Waste. There are a range of wastes associated with agriculture. As mentioned above, manures are a major waste problem for meat or dairy producers. However, food processing plants or operations often produce large amounts of organic waste that ranges from blood and offal from butchering sites to waste from vegetable and fruit processing. Because these sites often have homogenous waste over time, just as in industrial waste, interventions can be made to try to manage wastes appropriately so that they do not become a long-term pollution problem for a region.

Household Waste. Household waste is the waste that is produced day to day in a household and it consists of a highly complex mix of materials that makes processing it challenging. One of the most common components of household waste is organic material which includes food scraps, yard waste, and other biodegradable materials. Another component of household waste is recyclable materials. This typically includes paper, glass, and metals. Many communities require these materials to be separated from the waste so that it can be diverted to a recycling facility.

Composite wastes are also common in household waste. These are things that are made of multiple materials and include things like clothing, furniture, and Tetra Pak containers. These containers are unique in that they allow the shipment of milk and other products without pasteurization. Clean products like milk are superheated to kill any pathogens and are then placed within super clean containers. The containers are made of thin layers of plastic, paper, and aluminum and thus are challenging to recycle. Efforts are underway to expand the recycling of these containers, but most currently end up in the waste stream.

Household waste also includes some small amounts of hazardous waste (paints, herbicides, etc.), electronic waste, and medical waste.

How We Manage Waste

There are three main ways that we manage waste: landfills, burning, and recycling. Each of these has pluses and minuses to them that must be considered. It is also worth thinking through the impact of our bureaucratic waste system and how effective it is at promoting sustainability.

Sanitary Landfills. These facilities are sites where garbage is buried for long periods of time. Our modern sanitary landfills are lined to prevent leakage of liquids that could contaminate the subsurface. These liquids, called leachate, filter through landfill waste like water flowing through coffee grounds, picking up contaminants along the way. In addition, landfill layers are covered with regular layers of dirt, sometimes street sweeping debris, to try to avoid the waste blowing around and causing local litter problems. Commonly, methane builds up in landfills as organic materials decompose. As a result, methane is typically gathered and either burned as a waste or used for energy production. Groundwater off-site is monitored through wells and interventions are made if there appears to be any leakage from the facility. These facilities can be expensive to maintain and monitor and many parts of the world continue to use unsanitary landfills, or dumps, as their main form of waste management.

Some sanitary landfills specialize in particular types of wastes. For example, some may handle particular hazardous industrial wastes and others may handle specific agricultural wastes such as byproducts from meat processing. Each region of the world has its own unique garbage mix and sanitary landfills have to find ways to meet the needs of individuals and organizations in the region. Thus, the contents of landfills will vary from place to place. As a result, the environmental risk to the region's environment from landfills depends on the activities taking place within the garbage shed of the region. If a place has a number of industries or activities that create or use hazardous materials, it is likely that local landfills will have higher risks for contaminations than areas that do not have these industries or activities.

Because landfills are unpopular land uses in many communities, some have to ship their garbage great distances across state lines or even across national boundaries. New York City, for example, ships much of its garbage to other landfills around the state and to facilities as far away as Kentucky. Communities that take trash from other communities typically do so to create jobs for local residents. Sometimes these jobs come at environmental and public health costs.

Burning. We have burned our waste for thousands of years. However, burning garbage today can cause serious environmental problems. The smoke from garbage fires contains small particles that can enter the lungs to cause respiratory issues. Plus, our modern garbage contains hazardous materials that can be released into the environment upon burning and the remaining ash can concentrate pollutants like heavy metals. As a result, for many years, burning was not the favored approach to managing garbage. However, new smokestack technologies have made garbage burning in waste to energy facilities more environmentally sound.

Waste to energy facilities basically work like any power plant. They take fuel, in this case garbage, and burn it to create steam which turns a turbine when creates electricity. The smokestack technology filters the smoke to remove any hazardous materials. Overall, this approach is seen as environmentally sound because it removes the waste from landfills and replaces electricity that would have been derived from fossil fuels. It must be noted that there are still ash issues. While some of the ash is reused in a variety of products, much of it must be landfilled in specialty landfills.

Waste to energy plants are most commonly found in Europe, the United States, Japan, and China. These countries have advanced economies and produce a tremendous amount of garbage. Plus, many of them, particularly Japan and some countries in Europe, have limited energy supplies and electricity produced from waste to energy power plants is an important part of the energy supply. While there are limited numbers of waste to energy plants in other parts of the world, many countries in the Middle East, Africa, Asia, and South America are planning waste to energy facilities to deal with emerging waste management issues.

Recycling. Just as we have always burned garbage, we have always recycled our waste in some ways. For example, there is abundant evidence that prehistoric societies recycled stone tools to make new items and the ancient Egyptians even recycled coffins. Today, we look at recycling as a component of waste management in order to reduce the amount of materials entering landfills or waste to energy facilities, and as a way to reduce the need for consuming raw materials such as wood for paper or aluminum ore for aluminum cans. We are trying to both reduce waste and reduce the need to expand mines or cut forests.

Table 10.2 lists the major types of waste materials that can be recycled. Typically, we think of things like paper, aluminum, glass, and plastics as recyclable. However, most waste can be recycled or reused in some way. For example, we throw away tons of clothing in our world of fast fashion. Some of it can be diverted from the waste stream for reuse by someone who can find value from it. Similarly, sewage sludge can be diverted from some waste streams for agricultural fertilization. Indeed, in the last two decades, there has been a major push to focus on the development of a circular economy around waste. Organizations are examining their waste stream to find ways to divert their waste into useful materials. Thus, if you are home builder, you may have construction waste consisting of a

Table 10.2 List of major classes of recycled materials with examples

Type of material	Examples
Ceramics and rock-like materials	Dishes, brick, concrete, asphalt, plaster
Chemicals	Acids, fuels, lubricating oils, medicine
Glass	Bottles, window glass
Metals	Cans, construction metals, vehicles, screening, tools, appliances
Paper	Paper, newsprint, cardboard
Plant debris	Tree limbs, grass clippings, leaves
Plastics	Containers, packaging, household materials, tires
Putrescibles	Animal bodies, manure, agricultural waste, sewage sludge
Reusable materials	Working appliances, clothing, building materials
Soils	Excavation materials
Textiles	Clothing, drapery, furniture components
Wood	Lumber, pallets

range of building materials. By examining the contents of the waste, you can divert some of it for reuse and some of it can be made into useful products by other organizations. The key is to find users of the waste so that it is diverted from the waste stream.

Recycling is typically managed initially by the producers of waste who divert recycled materials from non-recyclable garbage. The separated recyclable materials are picked up by a waste hauler and typically taken to a recycling center. At these centers, the waste enters a complex separation line that includes both mechanical and human sorting processes. For example, many materials are separated by their weight using gravity or a burst of wind. Human sorting is often done for Tetra Pak cartons or some specialty plastics. At the end of the sorting line, some materials are not separated out and end up in a landfill. The collected materials, typically paper, plastic, and glass, are balled or packaged and stored for shipping to users of the materials.

While there are strong markets for some recycled goods, particularly paper and aluminum, there is only a modest market for plastics. Unfortunately, about 80% of the plastic produced in the United States is shipped outside of the country and about half of that is mishandled or ends up in the host country's waste stream. This is a problem in many developed nations: we want to recycle many things but there is a limited use for the recycled materials. Thus, there are certainly questions surrounding the ethics of recycling materials that are not really in demand as a product. While we can feel good about recycling plastics overall, the reality is that plastic recycling has limited value. The best we can do is to try to avoid the use of plastics in our day-to-day lives. Similar issues are associated with clothing. Retailers used to have distinct seasons for clothing. Now, with fast fashion, the idea of seasons is somewhat passé. Retailers that provide inexpensive clothing now constantly change clothing designs which draw in consumers to buy more and more clothing. As a result, we throw away huge amounts of clothing in our culture. While some of this clothing does end up in resale stores, the reality is that the stores are awash in used clothing because of the huge amount available at this moment in our history and they cannot sell all of the clothing that is donated. This shows us that not only do we have a waste problem, but we also have an overconsumption problem.

Stopping Consumerism

Of all of the ills that face Western society, consumerism is one of the worst. Our constant desire for more stuff causes problems all over the world. We live in a throwaway society that values instant gratification through purchasing over contentment with personal happiness. We are on a treadmill that focuses on getting more and more stuff. The television show *Hoarders* highlights the issue of too much "stuff" and how the stuff we own ends up owning us. In each episode, we meet someone who has started to accumulate so much stuff that their home is overtaken with garbage. Hoarding, of course, is a mental health issue. However, the hoarding problem is exacerbated by the range of easily available stuff we have in our society. Stuff is cheap and we can find inexpensive things to buy almost everywhere.

What makes our stuff so cheap in developed countries? Globalization is the engine driving the availability of cheap consumer goods. Companies use inexpensive labor and lax environmental regulations in developing countries to open up factories to manufacture stuff. Using the highly efficient global transportation system made up of container ships, rail, truck, and air freight, the stuff ends up in retail stores at prices far cheaper than if the stuff were manufactured in the West. This creates conditions where it is cheaper to purchase a range of new goods than to repair older stuff. Just a generation or two ago it was common to repair televisions, clothing, appliances, and a range of electronics. Now, it is often cheaper and more convenient to buy, say, a new dishwasher than repair one.

This attitude has changed generational attitudes around stuff. I remember working on a geologic project in the field in Arkansas years ago when my backpack ripped. I told a colleague who was a generation older than I was that I needed to stop and buy a new one. He asked me why I would buy a new one when I could just repair the one I had. He was right. My backpack could be easily repaired and, in that moment, I realized that my generation had a stuff problem. We grew up anticipating that we needed new things if there were small problems with our existing stuff. We did not value things in the same way as previous generations. Indeed, this issue has filtered down to generations that came after me and is spreading more and more to other cultures. We have a quest to purchase large volumes of consumer goods unlike any generation before us.

Our desire for constant gratification for consumer goods is driven, in part, by the modern marketing. A great example of how marketing drives our consumption can be seen in tech companies. It used to be that phones were more of an appliance. We had landline phones that were wired directly into our homes. The phones were not consumer goods. One purchased a phone, installed it, and it stayed as part of the house until the phone broke. Now, however, phones are consumer goods. Subtle annual changes in technology or style keep us coming back to purchase new ones every year or two. Our constant drive for new tech is changing our waste patterns and is driving decision-making around everything from mining operations to labor laws.

Tech is just one of the problems. Clothing, as previously mentioned, is also an issue. A couple of generations back, it was common to own a few high-quality pieces of clothes that were worn regularly and maintained and repaired as needed. Now, we have closets full of inexpensive clothing and we tend to throw away or donate clothing each year to make room for more. Even home furnishings and decorations have become part of the fast-fashion trend. Stores like Home Goods or online stores like Wayfair allow us to all be interior decorators who update our home's look with the season.

If you look closely, you can find overconsumption everywhere. Most of you reading this book probably have more stuff than you will ever truly need. This overconsumption, of course, leads to a lack of care about older things and as a result, we end up with lots of stuff entering our waste stream.

There is a growing social reaction to our overconsumption. For example, each year on Black Friday, the major shopping day in the United States on the day after Thanksgiving, a number of anti-consumerism groups celebrate Buy Nothing Day to help educate the public on issues of consumerism. During this time of year, many families go into significant debt to purchase stuff for the Christmas holidays. Buy Nothing Day advocates urge families to look carefully at their choices on Black Friday.

Another important anti-consumerist movement is called minimalism. The minimalist lifestyle is built around the idea that the stuff we own ends up owning us. Living with less actually frees us to live more interesting authentic lives. We are able to free up more money, spend less time taking care of stuff, and be more focused if we embrace minimalism. In recent years, the minimalist idea was highlighted by the popular documentary *The Minimalists*, which features two friends, Joshua Fields Millburn and Ryan Nicodemus, on their minimalist journey. They have also written books and host a popular podcast called *The Minimalists* (The Minimalists, 2020). They highlight that we should love people and use things and not the other way around. The focus of minimalism is on our relationships and our quality of life—not on what we own.

Minimalism has translated into a range of new lifestyle approaches. Some have downsized to live in small homes or apartments. Others have limited the amount of clothing and furniture they own. Some have gotten rid of cars and hobby equipment like sewing machines and saws. Others have stopped collecting stuff like books and knickknacks. Some have even gotten rid of personal items like photo albums, childhood remembrances like yearbooks, and cherished gifts. The overall point of minimalism is to have more time and space to live authentically at the moment.

It is also important to note that associated with minimalism is the waste-free lifestyle movement. Many people around the world are actively engaged in finding ways to cut down the waste that they produce so that their planetary burden is reduced significantly. They commit to limiting purchases to things that they need and to purchase things that do not come with packaging waste. For example, someone living a waste-free life would not purchase water or soda in a bottle. In addition, they would strive to buy materials in bulk using their own containers. They would not buy packaged oatmeal, for example, but purchase oatmeal in a store that sells in bulk and that would allow them to bring in their own containers for transporting the material. This commitment to a waste-free lifestyle does take planning but many have shown that they can live a modern lifestyle without producing significant amounts of waste. Many of these advocates take the lifestyle even further and eliminate even their recycling waste. Thus, they cut their paper, plastic, and metal use as much as possible. The waste-free lifestyle is not just about cutting landfill waste, but also about cutting the burden on recycling. Recycling, after all, is all about finding ways to deal with the waste that we produce.

Ways to Cut Waste

The previous sections reviewed a range of problems associated with the waste that we produce. Table 10.3 lists several ways that we can cut waste from our lives and live a more minimalist lifestyle. The easiest thing we can do to cut waste is to make sure we take advantage of whatever recycling opportunities exist within our communities. Most people that are serviced by some type of waste management company have at minimum plastic, paper, glass, and aluminum recycling available. In addition, it is relatively easy to start a food composting system if you have the space to compost by adding it to yard waste within composting bins. To deepen your commitment to cutting your waste, conduct a household garbage inventory. It is instructive to take a look at where your waste is coming from to understand how to cut it. For example, if you find you have

Table 10.3 Simple, hard/expensive, and innovative/life-changing approaches to reduce waste and live a minimalist lifestyle

	Ways to reduce waste and enhance recycling	Ways to live a more minimalist lifestyle
Simple	Do the basics of recycling in your community by separating recyclables from garbage. Start a food and yard waste composting system if you have the space available to you.	Celebrate "Buy Nothing Day" on Black Friday and reject the pressures of gift giving. Do an inventory of your stuff and try to eliminate clutter and excess stuff from your life.
Hard/ expensive	Conduct a household waste inventory to assess how you can reduce your waste.	Do not purchase things with excess packaging, plastics, or added waste burden. Evaluate how you spend time and assess how you can advance your cultural minimalism.
Innovative/ life-changing	Commit to living a zero-waste lifestyle.	Deepen your minimalism by getting rid of things in your home or office that have no meaning or use. Evaluate your housing and transportation choices.

mainly fast-food containers, you can make decisions to cut fast food from your life to cut the waste.

To take it even further, you can commit to living a zero-waste lifestyle which would require significant changes to the way that you shop. For example, many of us order a great deal of stuff online due to the convenience of online shopping. However, this comes at a great waste burden and a zero-waste lifestyle would cut online purchasing from your options. In addition, you would have to change the way that you bought food. For example, you would need to shop at a grocery store that allows bulk purchasing of materials using your own containers. While this may sound difficult, it is not. It is more awkward than anything since it requires a significant change of behavior of not only you, but others around you.

Of course, cutting waste is easier if you embrace the tenets of minimalism. The easiest way to embrace minimalism is to celebrate Buy Nothing Day instead of buying into the consumerism of Black Friday. In addition, it is relatively easy to get off the gift-giving treadmill in our society by giving cash or experiences as gifts. Another easy thing to do is to conduct an inventory of your stuff to evaluate what you own over what owns you and your time. To deepen your commitment to minimalism you can avoid purchasing things with excess packaging or waste burden. In addition, it is useful to evaluate how you spend time to evaluate how you can embrace a cultural minimalism. Many of us spend tons of time on the Internet or binge watching shows on our electronic devices. We can certainly find more productive and meaningful experiences for ourselves and our family. Finally, we can start to go further into minimalism by purging stuff that we own that has no direct use or meaning. In addition, we can evaluate our housing and transportation choices and make decisions that will lead us to live more sustainable authentic lives.

Reference

The Minimalists. (2020). *The Minimalists*. Retrieved December 5, 2020, from https://www.theminimalists.com

11

Saving Ecosystems

Introduction

The US state of Florida is an international tourist attraction that draws people to its sunny beaches and world-class theme parks. However, it was not always a draw. In the late nineteenth and early twentieth centuries, Florida was seen as a mosquito-infested backwater of the United States. It was dangerous because of its disease-carrying insects and because of the massive alligators that trawled the swampy waters. It also regularly experienced severe storms that destroyed ships and port communities. The soils of the state were not well-suited to agriculture and the extreme boom/bust seasonal variations of rainfall added to the problem. Plus, the state was dotted with wetlands and the whole southern half of the state was one vast wet grassland known as the Everglades. From the perspective of many nineteenth-century Floridians, it needed improvement. This was especially clear after a 1928 hurricane killed over 4000 people when a wall of water was blown out of Lake Okeechobee into the low-lying communities between the lake and the Everglades.

Prior to the hurricane, parts of the Everglades were drained to build sugar plantations and to make space for growing communities like

Miami. However, after the hurricane, engineers advocated stronger water control in the region to try to avoid flooding. A massive levee was built around Lake Okeechobee, the tenth largest lake in the world. In addition, a series of canals and drainage systems were constructed across the complex wetland system. In 1947, the US Congress formed the Central and Southern Florida Flood Control Project which greatly expanded on the series of canals and levees across the Everglades.

By the 1970s it was clear that the changes caused massive changes to the Everglades system. The natural flow of water from Lake Okeechobee was curtailed and ecosystems were under stress. Plus, the water entering the Everglades was different from that in the past. It contained huge amounts of nutrients running off of farm fields, residential lawns, and golf courses that sprung up in the Everglades drainage basin. Because the Everglades system is largely a vast delta from overland flow from central and south Florida, the wetland received lots of contaminated waters from all of the new urbanism that cropped up in the state in the twentieth century.

Since the 1970s, the state has been working at trying to improve a very damaged ecosystem. The author and activist, Marjory Stoneman Douglas, who passed away in 1998, was one of the leading advocates for the improvement of the Everglades. Her book, *The Everglades: River of Grass*, published in 1947, highlighted the beauty of the fragile ecosystem and the threat it was under and helped to awaken the public's mind about the importance of ecosystems in global cycles (Douglas, 1947). She, along with other mid- to late-twentieth-century advocates of ecosystem, such as ocean explorer Jacques Cousteau, helped us all understand how we were profoundly impacting ecosystems everywhere.

This chapter focuses on the range of ecosystems around the world and highlights some specific examples about how we are doing profound damage to them. The chapter concludes by reviewing how we can work to preserve and protect ecosystems around the world.

Ecosystems

Ecosystems are simply assemblages of organisms and their environment. They are systems in that the organisms and their environment evolve over many years. They influence each other. For example, a prairie environment contains grasses that help to create particular soils that favor prairies. As grasses grow, their roots penetrate deep into the earth. When the grasses die, the roots decompose and help to fertilize and loosen the soil to promote better grass growth. In another example, there are pine trees that evolved with regular brush fires in naturally fire-prone areas. Seeds in pine cones release only at extremely high temperatures. What is important to stress is that ecosystems evolve slowly over time and become rather stable. While all ecosystems are changing in some way at all times, most natural ecosystem change happens slowly.

There are two broad classes of ecosystems: terrestrial and aquatic. Each of these can be further broken down into smaller units. A review of the main ecosystem types is listed in Table 11.1.

Terrestrial ecosystems include forests, grasslands, tundra, and deserts. Each of these can be further subdivided. The forest ecosystems include tropical forests, temperate forests, boreal forests, and savannas. Each of these, of course, can be further subdivided into specific types of forests such as a tropical rain forest. Grassland ecosystems include tropical, temperate, flooded grasslands, alpine, and desert grasslands. The tundra ecosystem is divided regionally into Arctic, Antarctic, and alpine tundra. Finally, the desert ecosystem is divided into dry, semiarid, coastal, and cold deserts.

Aquatic ecosystems are divided into two broad classes. Marine ecosystems are very diverse and include a range of coastal and near coastal ecosystems including estuaries, beaches, salt marshes, mangrove forests, and coral reefs. In addition, the marine ecosystem includes open ocean areas and deep ocean areas.

As can be seen by the classes of ecosystems, they are highly dependent on climate and situation. For example, the forest ecosystems are clearly based on their climate. However, some ecosystems, such as alpine grasslands, are clearly based on their situation within the landscape. Regardless,

Table 11.1 Major types of wetlands

Terrestrial	Forest	Tropical forest
		Temperate forest
		Boreal forest
		Savanna
	Grassland	Tropical grasslands
		Temperate grasslands
		Flooded grasslands
		Alpine grasslands
		Desert grasslands
	Tundra	Arctic tundra
		Antarctic tundra
		Alpine tundra
	Deserts	Dry deserts
		Semiarid deserts
		Coastal deserts
		Cold deserts
Aquatic	Marine	Estuaries
		Salt marshes
		Mangrove forests
		Coral reefs
		Beaches
		Open ocean
		Deep sea ocean
	Freshwater	Lakes and ponds
		Rivers
		Wetlands

there is a distinct link between type of ecosystem and their broader environment.

Humans and Ecosystems

Humans are natural beings, and thus like other organisms, we alter ecosystems through our existence. There are many examples of humans altering ecosystems throughout time (Martin, 2005). For example, prehistoric peoples in North America changed environments through their hunting practices. They were partially responsible for the extinction of some of the megafauna across the continent. In addition, deforestation wiped out some of the forest ecosystems in the Mediterranean region thousands of years ago.

More recently our imprint on ecosystems has grown and there is evidence of human alterations of ecosystems all over the world (Ramankutty & others, 2002). In some cases, the changes are subtle. For example, we can find changes in soil or atmospheric chemistry in uninhabited places in the Arctic as a result of the atmospheric deposition of low levels of pollutants like heavy metals or nuclear fallout. We can also find plastic in most areas of the oceans. These changes to ecosystems may seem small, but they can have important impacts. For example, most seabirds have ingested plastics in some way and there is real concern over how our relatively recent cultural embrace of plastics and the resultant waste it creates will impact our planet.

In some cases, ecosystems help to process many of the problems that we impose upon them with limited impacts. One of the most important of these ecosystem services is the absorption of pollutants—particularly nutrient pollutants. Indeed, ecologists can place monetary value on the costs of losing ecosystem services if ecosystems are destroyed.

In some cases, the impact of our activities on the planet has totally changed systems (Ekstrom & others, 2015). We have cut forests, drained swamps, and watered deserts. We have destroyed coral reefs and polluted estuaries. This does not mean that ecosystems are not present, but they are altered and transformed into something that was not there before (Raptis & others, 2016). As a result, the plant and animal species that once existed in a place may end up threatened, endangered, or extinct and exotic plant and animal species can move in (Henry & others, 2012).

An endangered species is a plant or animal that is likely to become extinct without some type of intervention. Exotic species are those species that are not native to a region but take the space of native species in an ecosystem. It is worth returning to the Florida Everglades example to review these three classes of species. The Florida panther is a type of North American cougar that lives in isolated habitat in south Florida. It once roamed across wide areas of the southeastern United States and probably had connections with other cougar species across the continent. Today, however, there are only around 200 remaining in the wild. They have been killed by farmers and run over by cars. Because they need a large habitat and traverse large distances, they have been hemmed in by extensive urbanization across the Florida peninsula. Because the Florida panther is an officially endangered species, a number of interventions have been put into place to try to protect it. For example, the state has built tunnels under roadways to allow panthers to get from one habitat to another without the risk of getting hit by a car. In addition, the state and federal governments have tried to connect habitats by purchasing private lands and allowing them to revert to natural habitats.

Threatened species are those that are at risk of declining to the point that they are endangered. The Florida manatee was one on the endangered species list but was moved to the threatened list in 2017. The manatee, sometimes called a sea cow, is a large species that lives in warmer coastal waters of Florida. It can also enter estuarine and riverine waters and they are also known to cross the Florida peninsula through canals via Lake Okeechobee. They were once hunted for food, but now their main threats are boats. About a third of all manatee deaths are associated with boat motor strikes. A range of protections have helped the species recover across the state. Some coastal areas have been set aside as refuges where boats are not allowed and there are other areas where intensive boater education has helped create awareness that boaters need to keep watch for the large creatures in some key areas.

Exotic species are those species that do not belong in a particular ecosystem. They often change ecosystems negatively by displacing native species and transforming ecosystems into something entirely different from what evolved over hundreds, if not thousands, of years. One of the most harmful species to invade the Florida Everglades is the Burmese python. This animal, which is native to Asia, was a popular pet in Miami and some, of course, escaped into the wilds of the vast peninsula. Now, there are likely hundreds of thousands of the constrictors in the Everglades and they have fundamentally changed the ecosystem. The pythons have been eating their way across the Everglades and they have decimated the population of mid-sized mammals such as foxes, rabbits, bobcats, raccoons, and opossums. Officials in Florida have reacted to the problems caused by pythons by enacting special hunting seasons and by trying to more carefully regulate the exotic pet market.

The Florida Everglades is but one example of a place where endangered, threatened, and exotic species are of concern. In all corners of the world, there are issues and human activity is responsible for all of them. However, there are some specific areas, called urban ecosystems, which have been completely transformed and are unrecognizable as anything approaching natural habitats that existed prior to urbanization.

Urban Ecosystems

While it may seem counterintuitive, urban places are ecosystems—they are assemblages of organisms and their environment (Fig. 11.1). Cities are hardly places where one would look for natural systems, but they are worth examining from an ecosystem lens. Urban areas have social and economic systems that impact local and surrounding environments.

Fig. 11.1 Florida's wetland ecosystems contain a tremendous amount of biodiversity, even though they have been heavily altered by human activity

Geographically, these social and economic systems vary considerably from place to place, and as a result, urban areas have all developed unique relationships with the environment. For example, New York City, which is largely an island city surrounded by water, has dense development with few natural areas outside of parks like Central Park. London, which is a more sprawling city built along a river, has more park-like land in backyard gardens along with parks. The Los Angeles, Houston, or Cairo regions, with their vast network of roadways, have transformed huge tracts of land into urban environments.

The relatively new field of urban ecology has emerged in the last 50 years or so to study the relationships of the social and economic systems in cities and the environment. The field of urban ecology is rather broad and encompasses a range of topics. Often, urban ecologists study species in urban areas. They could, for example, study pesky rat populations to better understand how they can be brought under control if they are causing problems. Or, they could study native plant species and how they are able to survive in cities. The field is often very applied in that urban ecologists are trying to solve problems or make improvements.

An example of how urban ecologists work to make improvements could be seen in the work of urban forest programs. With all of the energy expended in transportation, heating, and cooling in cities, and with all of the urban hard surfaces that absorb solar radiation, a new phenomenon has emerged called an urban heat island. Temperatures in many cities are slightly higher than those in surrounding rural areas. The excess heat can cause serious problems for cities such as increased costs for cooling and health problems for residents. In addition, the added heat transforms urban ecosystems by changing the overall climate zone of cities. To counteract this problem, urban ecologists in some cities have developed comprehensive urban forestry programs that promote cooling and the absorption of the heat energy by vegetation. Urban forests can limit the impacts of the urban heat island effect considerably.

Urban ecologists are often at the forefront of trying to limit the broader impacts of urbanization on surrounding ecosystems. For example, in Milan, Italy, urban ecologists seek to develop urban boundary areas and to prevent damage to the environment from Milan's considerable industrial footprint. In San Francisco, urban ecologists have worked hard to limit destruction of the coastline and advanced efforts at restoration of coastal wetland ecosystems. In Milwaukee, urban ecologists have built ecosystem connectivity through the park system. And throughout Asia, urban ecologists are seeking to restore riverine systems in cities to try to prevent a range of pollution problems.

Unique Ecosystems Under Threat

Wetlands. When one hears the term wetlands, one may think of a dank, dangerous smelly area (Fig. 11.2). While some wetlands can be dank and smelly, and even dangerous, they vary quite widely and cannot be easily classified except for the fact that they are ecosystems that are permanently or seasonally flooded. They are broadly classified by the type of vegetation found within them. Forested wetlands are swamps and grassy wetlands

Fig. 11.2 Urban ecosystems, such as this one on my campus in DeKalb, Illinois, are home to a fascinating array of plants and animals—and humans

are marshes. Acidic mossy wetlands are called bogs. Swamps are further classified by the type of trees present in the wetland. The terms mangrove swamp (tropical coastal wetlands), cypress swamp, and spruce swamp are common names. Marshes are classified by water type: saltwater marsh, freshwater marsh.

Wetlands are often transitional landscapes between open water and upland environments. The vast coastal marshes of the southeastern United States, for example, are a transition between the agricultural coastal plain and the Atlantic Ocean. Some wetlands, such as the cypress domes found within peninsular Florida, are small, isolated features not connected to open water at all. The moisture comes from groundwater and some of the domes are transitional landscape features as ponds or lakes fill in over long periods of time with sediment and organic matter.

Wetlands have been under threat in the twentieth and twenty-first centuries throughout the world. Around 35% of all the world's wetlands have disappeared in the last century (Davidson, 2014). The distribution of this loss is not uniform. For example, around half of all wetlands in the United States have been destroyed. Most of the destruction is a result of urbanization and agriculture.

There are a range of efforts underway to protect wetlands. For example, some governments have set aside large areas of wetlands as parks or preserves. Everglades National Park is an example of such a preserve. In addition, many governments have made destroying wetlands illegal or in some other way regulate the development of wetlands. One initiative that has garnered considerable support in recent years is the notion of wetlands banking. What this means is that a developer could pay to destroy a small wetland in some area with the money going to support the purchase of wetlands in other areas. Typically, the funds go to support the purchase of large areas of land in order to preserve large swaths of these fragile ecosystems.

Coral Reefs. Coral reefs are large assemblages of coral and the other animals that have evolved with the coral. What is fascinating about coral reefs is that the coral polyps build up the ecosystem through the deposition of calcium carbonate to build up the rock in which the coral live. Coral reefs are found extensively in shallow warm tropical areas where the water is clear and where there is some active movement of water. Expansive coral reefs are found in the Caribbean, off the coast of Australia (The Great Barrier Reef), and parts of Southern Asia.

Coral ecosystems are declining rapidly. They are damaged by rising sea temperatures, ocean acidification, nutrient pollution, and runoff from overdevelopment. Sunscreen has also been shown to cause damage to them. Coral bleaching, one of the most significant problems for coral reefs, is caused by slight increases in ocean temperature. One of the most significant coral reefs in the world, The Great Barrier Reef off the west coast of Australia, is in significant decline. About half of the coral in the reef has died as a result of coral bleaching in the last few decades.

There are efforts underway to try to preserve and protect coral. Many countries have developed marine sanctuaries that allow limited boat access to the reefs. In addition, there are a variety of coral restoration efforts underway. Some of these involve creating new coral colonies in places where none existed before. In other places, coral farming is underway to start coral colonies for transplantation in areas where colonies are damaged. Coastal communities have also developed policies to try to prevent coral damage by instituting regulations such as prohibiting the use of sunscreen if swimming in coastal waters. Even with all of these efforts, however, coral reefs continue to decline and there is concern as to whether they will be able to survive as our climate changes and as we continue to alter our coastlines.

Tropical Rainforests. The tropics are roughly defined as the areas between the latitudes of 23.5 degrees north and south. These equatorial regions have little temperature variability and the climate is warm and humid with intense rains. Because of these conditions, there is tremendous biological productivity. Plants grow quickly and trees can reach tremendous heights. Tropical rainforests are one of the most characteristic ecosystems in these areas.

The tropical rainforest ecosystem has unique soil characteristics. Even though the ecosystem is highly productive, soils under the forest tend to be poor. Due to all of the rain, nutrients wash through the soils quickly. In addition, the soils are heavily oxidized and have a deep red color. The biological productivity of the ecosystem is dependent on a process called nutrient recycling. In this system, plants and animals in tropical rainforests release their stored nutrients upon death and subsequent decay and they are quickly picked up by plants through their extensive root systems. Thus, if vegetation is removed from the landscape, say from extensive deforestation, the soils quickly lose their ability to support a productive ecosystem.

Indigenous people living in tropical rainforests developed a distinct type of agriculture that is in tune with this local ecology called slash and burn agriculture or milpa agriculture. They slash away small areas of the rainforest where they plant small gardens that will feed their family or their community for a year or two. When the productivity starts to wane, they move on to another site and let the forest take the garden over to enhance the nutrient content of the soil. All over the world, indigenous people living in tropical rainforests have relied on this type of agriculture for centuries.

Today, tropical rainforests are under threat. There are three broad areas of tropical rainforests in the world. The largest of these is the Amazon in South America. There are also extensive forests in Central Africa and in Southeast Asia. The main reason all of these systems are under threat is due to agricultural development. In Southeast Asia, the main threat is from the expansion of palm oil plantations. Palm oil has become a popular building block in the global food production chain. It is also used in products like lipstick and biofuel. As a result of the strong international demand for palm oil, tropical countries like Malaysia and Indonesia have very rapidly developed palm oil plantations at the expense of rainforests.

The tropical rainforests of Africa have declined for somewhat different reasons. The growth of agriculture is certainly one piece of the story, however, overall population expansion in the rainforest has segmented it significantly. Thus, while local people still practice slash and burn agriculture, there just isn't that much forest left. Mining is another issue in the region.

The South American Amazon rainforest has been under attack for generations, but its destruction has accelerated in the last few years. While the rainforest extends into multiple countries, most of it is within Brazil and it is worth taking a look at their policies to understand why the ecosystem is in the midst of destruction. For generations, the Amazon was open to development. However, due to the rough tropical conditions and due to the prevalence of tropical diseases, it was not seen as a desirable location for human settlement and most of the population of Brazil lived near the Atlantic coastline. Some inroads were made in the early and mid-twentieth century and cattle ranching flourished in some locations. By the second half of the twentieth century, there was more pressure from rural communities to further develop the Amazon for agriculture and forestry.

This, of course, led to conflict with indigenous peoples who lived in the rainforest for hundreds of years. As forests were cut and as ranching expanded, they were forced further inland. Ranchers killed or otherwise displaced native Brazilians in order to further their economic gains across the Basin. Under the Presidency of Fernando Henrique Cardoso, who was elected to the Brazilian presidency in 1995, steps were taken to try to protect indigenous peoples in the Amazon by setting aside reservations. These efforts were more or less deepened by subsequent presidents up through 2019 with the election of Jair Bolsonaro. For example, former president Dilma Rousseff greatly expanded indigenous rights, but also supported several hydroelectric development projects in native people's lands.

The election of Jair Bolsonaro changed everything in Brazil. He ran on a largely populist platform that included the development of the Amazon rainforest as a central theme of his campaign. Many ranchers and farmers felt that their economic potential was hindered by national policy toward protecting the forest and its inhabitants. When Bolsonaro won election in 2019, he instituted several policies that made the development of the rainforest inevitable. Since his election, large swaths of the tropical rainforest have been destroyed. Environmental activists who have tried to protect the forests have been murdered in cold blood.

What is so significant about the loss of rainforests around the world is that we are losing some of the world's most biologically diverse landscapes. For example, the Amazon rainforest contains about one-third of all known tree species and you can easily find hundreds of different species of plants and animals in an acre of land. The extinction rate of tropical plants and animals is high. It is hard to see much hope for these precious landscapes since their destruction is increasing even as these words are written.

Permafrost Ecosystems. The landscapes of the far north and south are quite different from the tropical forest landscapes discussed in the previous section. Permafrost ecosystems are those ecosystems that evolved in places where permafrost is present in the subsurface. These places are quite cold most of the year. During the summer, the surface thaws, but parts of the subsurface never really thaw out. As a result, these permafrost regions have short, intense growing seasons. What is interesting about these permafrost landscapes is that many of them are wetlands. Even though precipitation levels are relatively low, the permafrost inhibits drainage and moisture can collect at the surface to create marshy (grasses), swampy (trees), and boggy (mosses) landscapes. Because so many of these permafrost systems are wet, huge amounts of dead vegetation are stored in anaerobic condition. For example, huge amounts of peat can build up in bogs in permafrost areas.

Permafrost ecosystems are highly sensitive to development and climate change. Subtle changes in temperature can transform the local ecosystem and melt the permafrost. When this happens, the landscape can drain and the former anaerobic conditions become oxidized and the stored organic matter can decay. It does not take much change to transform these places and scientists are already seeing ecosystem changes in the Arctic.

Because these permafrost zones are so sensitive to change, environmental policy makers have worked hard to protect them from development. For example, the U.S. has for decades set aside vast areas of the Alaskan Arctic as a preserve. Even though there is a great deal of oil and natural gas in the area, it was thought too sensitive for development. However, former President Trump, in August of 2020, changed this conservation course and sought to open up the lands for development. With his defeat at the polls in November of the same year it is doubtful that this initiative will proceed. However, this situation shows that there are intense development pressures on the permafrost regions of the world.

How You Can Protect Ecosystems

There are several things that you can do to try to preserve and protect ecosystems (Table 11.2). At the most basic level, you can avoid the release of chemicals, such as fertilizers, pesticides, and herbicides that do damage

Table 11.2 Simple, hard/expensive, and innovative/life-changing approaches to protecting and preserving ecosystems

	Ways to protect and preserve ecosystems
Simple	Avoid releasing basic pollutants (such as fertilizers) that can alter ecosystems
	Visit local parks to appreciate nature
Hard/expensive	Landscape only with local native vegetation
	Get involved with land and ecosystem preservation in your community
Innovative/life-changing	Commit to living a zero-waste lifestyle in harmony with your ecosystem
	Advocate internationally for preservation of ecosystems

to ecosystems. These chemicals cause innumerable problems from eutrophication to the formation of dead zones in coastal waters. The world's honey bees are declining through a problem called colony collapse disorder which may be caused by pesticides. It is also important to spend time in nature to appreciate it. Thus, an easy thing that any of us can do is to spend time appreciating our local ecosystems.

To take it one step further, you can commit to altering the landscaping around your home with local native vegetation. This will help restore some of the damage that is done to local ecosystems due to human activity. The landscaping will help to support local insects, mammals, amphibians, reptiles, and birds by providing food sources and shelter. If you live in an apartment, you can advocate locally in your community for landscaping public space with local vegetation. Another thing you can do is to advocate for the preservation and protection of ecosystems in your community and region. There are a number of ways to get involved. For example, most local governments have some park or preservation organization that protects land. In addition, many non-profit organizations, such as land trusts, work to purchase and preserve ecosystems.

To deepen your commitment to ecosystem preservation, you can live, as was suggested in previous chapters, a waste-free life in harmony with your ecosystem. Most of you reading this book consume much more stuff than the average person. Our consumption patterns drive the destruction of ecosystems. By inventorying our consumption and its impacts, we can make decisions as to which of our consumptive patterns are harmful to ecosystems. For example, coffee drinkers can commit to only drinking coffee certified by the Rainforest Alliance, a group that certifies that coffee plants are grown in harmony with nature. In addition, you can get involved either through action or through financial support with organizations that are working on large ecosystem protection projects. For example, there are many groups working to protect the large mammals of Africa, the rainforests of Brazil, and the coral reefs in the Caribbean and Australia. The bottom line is that ecosystems are in serious decline around the world and we all need to do as much as we can to preserve and protect them.

References

Davidson, N. C. (2014). How much wetland has the world lost? Long-term and recent trends in global wetland area. *Marine and Freshwater Research, 65*, 934–941.

Douglas, M. S. (1947). *The Everglades, river of grass*. Rinehart & Company.

Ekstrom, J. A., Suatoni, L., Cooley, S. R., Pendleton, L. H., Waldbusser, G. G., Cinner, J. E., Ritter, J., Langdon, C., van Hooidonk, R., Gledhill, D., Wellman, K., Beck, M. W., Brander, L. M., Rittschof, D., Doherty, C., Edwards, P. E. T., & Portela, R. (2015). Vulnerability and adaptation of US shellfisheries to ocean acidification. *Nature Climate Change, 5*, 207–214.

Henry, M., Beguin, M., Requier, F., Rollin, O., & Odoux, J. (2012). A common pesticide decreases foraging success and survival in honey bees. *Science, 226*, 348–350.

Martin, P. S. (2005). *Twilight of the mammoths: Ice age extinctions and the rewilding of America*. University of California Press.

Ramankutty, N., Foley, J. A., & Olejniczak, N. J. (2002). People on the land: Changes in population and croplands during the 20th century. *Ambio: A Journal of the Human Environment, 31*, 251–257.

Raptis, C. E., van Vliet, M. T. H., & Pfister, S. (2016). Global thermal pollution of rivers from thermoelectric power plants. *Environmental Research Letters, 11*. Retrieved June 1, 2019, from https://iopscience.iop.org/article/10.1088/1748-9326/11/10/104011

Part IV

Building Just and Equitable Economic and Social Systems

The murder of George Floyd in the summer of 2020 highlighted inequities in social justice in the United States and sparked international conversations around social justice and equality. We know that environmental challenges are not felt equitably throughout the world and that the decisions that impact the environment are not made by the people who are most affected. The following chapters provide a review of several issues related to equitable economic and social systems within the context of sustainability and they also provide ways to make improvements. Chapter 12 focuses on social issues and highlights several issues associated with environmental racism and human rights. Chapter 13 discusses how to transform economic systems so they focus more on community sustainability. Travel and tourism are highlighted in Chap. 14. Finally, the book ends with Chap. 15 on the theme of education.

12

Building a Just and Sustainable Society

Introduction

In the first two-thirds of the twentieth century, lead was used widely in a range of products such as paint and gasoline. It was used as an additive to paint to make it more durable and shinier. It was added to gasoline as an anti-knock agent. By the 1960s lead was not only used widely, it was also present as a pollutant. It was present in soil due to chipping and flaking of paint and due to tailpipe emissions from burning lead-containing gasoline. It was also present in dust in homes due to aerosol deposition of tailpipe emissions and from flaking paint. While lead was widely known to cause health problems, even dating back to preindustrial times (Riva & others, 2012), leaders in the twentieth century were willing to expose people to lead because they felt the products were worth the risk (Markowitz & Rosner, 2013).

Unfortunately, by the 1960s, lead pollution started to emerge as a public health problem. Lead is not used for any metabolic process and thus its ingestion or inhalation into the body can lead to significant problems. One of the most problematic issues is that it is absorbed by children much more than adults. Thus, children are far more susceptible to lead poisoning. Sadly, lead can impact cognitive function and development. As a result, lead pollution can lead to widespread learning difficulties in children. Some have even equated lead pollution with increases of violence (Mielke & Zahran, 2012).

It is important to note, within the context of this chapter, that lead pollution is not distributed evenly across the planet. In the United States in the 1960s and 1970s, most of the lead pollution was located in older neighborhoods of cities. This makes some sense since older homes typically have multiple applications of paint. In addition, older parts of cities tend to have dense traffic patterns and thus more tailpipe emissions. Plus, these older parts of cities would have had the dense traffic patterns for years compared to newer suburbs.

During the same period, American cities were in transition. Many white Americans were leaving the cities for the suburbs, usually leaving large populations of Black and Latinx citizens behind. What this means in the context of lead poisoning is that Black and Latinx children were disproportionately exposed to lead pollution for decades. Two trajectories for children emerged in the United States during the 1960s and 1970s that exposed real public health divides. White children were far less likely to have lead poisoning than Black or Latinx children. Because the families of Black and Latinx children tended to be in poorer communities and families, community health experts and families typically did not have the financial resources to mitigate the issues by removing lead from homes and soil, moving, or conducting health interventions (Newsome & others, 1997).

While lead was eventually banned in most products in the United States by 1979, manufacturers of lead-based paint and leaded gasoline knew the public health problems associated with their products. Instead of working to try to educate the public about them or trying to eliminate lead from their products, they denied that the problem existed or denied the magnitude of the public health challenges. This issue is broadly discussed in the groundbreaking book, *Deceit and Denial: The Deadly Politics of Industrial Pollution* by Gerald Markowitz and David Rosner (2013), which also discusses similar issues with the petroleum industry. This situation highlights how societies create environmental inequalities, often for financial gain.

This chapter seeks to examine efforts that are underway to build just societies around the world and how we can individually work to advance social justice and sustainability in our lives and in our communities (Fig. 12.1). After an examination of the meaning of human rights, the chapter moves toward more focused themes such as environmental racism and environmental justice. Some specific examples are provided prior to concluding with some steps you can take to advance sustainability and equity around you.

Fig. 12.1 The world has struggled with maintaining human rights around the world. What can you do in your community to try to make sure that human rights are protected?

The Universal Declaration of Human Rights

Human rights are those rights that we all hold in common. A variety of documents, such as Thomas Paine's *The Rights of Man* or the US Declaration of Independence, have tried imperfectly to provide a broader definition of what we today call human rights. However, it was not until 1947 that the world came together to create an international statement that defines human rights across the planet. The document, known as the Universal Declaration of Human Rights, was adopted by the United Nations shortly after the end of World War II (United Nations, 2020). Informed heavily by the atrocities conducted by the Nazis, the Declaration makes clear that all people have basic human rights.

The Declaration is made up of 30 different articles (Table 12.1) that provide a worldwide foundational understanding of human rights. They invoke important ideas about basic personal freedom, equal protection under the law, privacy, freedom of movement, workers' rights, marriage and family rights, property rights, access to representational government, freedom of expression and assembly, and rights to education and artistic expression. The Declaration draws together a number of important human rights threads that emerged decades and centuries before into a single holistic document that provides a framework for global cooperation toward improving the human condition. Each December 10th, the world pauses to celebrate Human Rights Day to draw attention to the 30 articles.

Over the years since 1947, there have been numerous failures of human rights that have led to a range of human suffering. Since the genocide of the 1940s in Europe, several other genocidal events have occurred in the Balkans, in Central Africa, and in Cambodia, to name a few. In addition, wars and forced migrations have added to global instability. The world continues to struggle with human trafficking and advanced technology has aided repression of truth and a free press around the world. Thus, even though we have the Universal Declaration of Human Rights, it is an ideal for which the whole world continues to strive to achieve. There are always demagogues who try to subvert the rights of others to achieve

Table 12.1 Articles of the Declaration of Human Rights

Article 1. All human beings are born free and equal in dignity and rights. They are endowed with reason and conscience and should act towards one another in a spirit of brotherhood.

Article 2. Everyone is entitled to all the rights and freedoms set forth in this Declaration, without distinction of any kind, such as race, color, sex, language, religion, political or other opinion, national or social origin, property, birth or other status. Furthermore, no distinction shall be made on the basis of the political, jurisdictional or international status of the country or territory to which a person belongs, whether it be independent, trust, non-self-governing or under any other limitation of sovereignty.

Article 3. Everyone has the right to life, liberty and security of person.

Article 4. No one shall be held in slavery or servitude; slavery and the slave trade shall be prohibited in all their forms.

Article 5. No one shall be subjected to torture or to cruel, inhuman or degrading treatment or punishment.

Article 6. Everyone has the right to recognition everywhere as a person before the law.

Article 7. All are equal before the law and are entitled without any discrimination to equal protection of the law. All are entitled to equal protection against any discrimination in violation of this Declaration and against any incitement to such discrimination.

Article 8. Everyone has the right to an effective remedy by the competent national tribunals for acts violating the fundamental rights granted him by the constitution or by law.

Article 9. No one shall be subjected to arbitrary arrest, detention or exile.

Article 10. Everyone is entitled in full equality to a fair and public hearing by an independent and impartial tribunal, in determination of his rights and obligations and of any criminal charge against him.

Article 11. (1) Everyone charged with a penal offence has the right to be presumed innocent until proved guilty according to law in a public trial at which he has had all the guarantees necessary for his defense. (2) No one shall be held guilty of any penal offence on account of any act or omission which did not constitute a penal offence, under national or international law, at the time when it was committed. Nor shall a heavier penalty be imposed than the one that was applicable at the time the penal offence was committed.

Article 12. No one shall be subjected to arbitrary interference with his privacy, family, home or correspondence, nor to attacks upon his honor and reputation. Everyone has the right to the protection of the law against such interference or attacks.

Article 13. (1) Everyone has the right to freedom of movement and residence within the borders of each state. (2) Everyone has the right to leave any country, including his own, and to return to his country.

(continued)

Table 12.1 (continued)

Article 14. (1) Everyone has the right to seek and to enjoy in other countries asylum from persecution. (2) This right may not be invoked in the case of prosecutions genuinely arising from non-political crimes or from acts contrary to the purposes and principles of the United Nations.

Article 15. (1) Everyone has the right to a nationality. (2) No one shall be arbitrarily deprived of his nationality nor denied the right to change his nationality.

Article 16. (1) Men and women of full age, without any limitation due to race, nationality or religion, have the right to marry and to found a family. They are entitled to equal rights as to marriage, during marriage and at its dissolution. (2) Marriage shall be entered into only with the free and full consent of the intending spouses. (3) The family is the natural and fundamental group unit of society and is entitled to protection by society and the State.

Article 17. (1) Everyone has the right to own property alone as well as in association with others. (2) No one shall be arbitrarily deprived of his property.

Article 18. Everyone has the right to freedom of thought, conscience and religion; this right includes freedom to change his religion or belief, and freedom, either alone or in community with others and in public or private, to manifest his religion or belief in teaching, practice, worship and observance.

Article 19. Everyone has the right to freedom of opinion and expression; this right includes freedom to hold opinions without interference and to seek, receive and impart information and ideas through any media and regardless of frontiers.

Article 20. (1) Everyone has the right to freedom of peaceful assembly and association. (2) No one may be compelled to belong to an association.

Article 21. (1) Everyone has the right to take part in the government of his country, directly or through freely chosen representatives. (2) Everyone has the right of equal access to public service in his country. (3) The will of the people shall be the basis of the authority of government; this will shall be expressed in periodic and genuine elections which shall be by universal and equal suffrage and shall be held by secret vote or by equivalent free voting procedures.

Article 22. Everyone, as a member of society, has the right to social security and is entitled to realization, through national effort and international co-operation and in accordance with the organization and resources of each State, of the economic, social and cultural rights indispensable for his dignity and the free development of his personality.

(continued)

Table 12.1 (continued)

Article 23. (1) Everyone has the right to work, to free choice of employment, to just and favorable conditions of work and to protection against unemployment. (2) Everyone, without any discrimination, has the right to equal pay for equal work. (3) Everyone who works has the right to just and favorable remuneration ensuring for himself and his family an existence worthy of human dignity, and supplemented, if necessary, by other means of social protection. (4) Everyone has the right to form and to join trade unions for the protection of his interests.

Article 24. Everyone has the right to rest and leisure, including reasonable limitation of working hours and periodic holidays with pay.

Article 25. (1) Everyone has the right to a standard of living adequate for the health and well-being of himself and of his family, including food, clothing, housing and medical care and necessary social services, and the right to security in the event of unemployment, sickness, disability, widowhood, old age or other lack of livelihood in circumstances beyond his control. (2) Motherhood and childhood are entitled to special care and assistance. All children, whether born in or out of wedlock, shall enjoy the same social protection.

Article 26. (1) Everyone has the right to education. Education shall be free, at least in the elementary and fundamental stages. Elementary education shall be compulsory. Technical and professional education shall be made generally available and higher education shall be equally accessible to all on the basis of merit. (2) Education shall be directed to the full development of the human personality and to the strengthening of respect for human rights and fundamental freedoms. It shall promote understanding, tolerance and friendship among all nations, racial or religious groups, and shall further the activities of the United Nations for the maintenance of peace. (3) Parents have a prior right to choose the kind of education that shall be given to their children.

Article 27. (1) Everyone has the right freely to participate in the cultural life of the community, to enjoy the arts and to share in scientific advancement and its benefits. (2) Everyone has the right to the protection of the moral and material interests resulting from any scientific, literary or artistic production of which he is the author.

Article 28. Everyone is entitled to a social and international order in which the rights and freedoms set forth in this Declaration can be fully realized.

(continued)

Table 12.1 (continued)

Article 29. (1) Everyone has duties to the community in which alone the free and full development of his personality is possible. (2) In the exercise of his rights and freedoms, everyone shall be subject only to such limitations as are determined by law solely for the purpose of securing due recognition and respect for the rights and freedoms of others and of meeting the just requirements of morality, public order and the general welfare in a democratic society. (3) These rights and freedoms may in no case be exercised contrary to the purposes and principles of the United Nations.

Article 30. Nothing in this Declaration may be interpreted as implying for any State, group or person any right to engage in any activity or to perform any act aimed at the destruction of any rights and freedoms set forth herein.

their own ends for their own personal gain or for the benefit of their community or nation over another.

Since 1947, when the Declaration was first published, a range of emerging issues has highlighted the evolution of human rights in the world's consciousness. For example, there has been significant advancement of LGBTQ rights in the last two decades and many countries now provide rights for same-sex marriages. In addition, some communities are articulating particular rights to animals, particularly pets which are often described in this context as companion animals.

The idea of human rights also intersects with rights to access various resources. For example, water advocates worry that privatizing water delivery limits the public's access to fresh drinking water. Also, what rights do citizens have to clean air or to natural outdoor space? Some of these rights issues can be contentious. For example, someone who paid millions of dollars to build a beach home may not believe that the public should have access to the beach in front of the home. As populations have increased and as income inequality has increased across the planet, the idea of access to natural spaces and to resources is growing in significance.

Environmental Racism

As noted in Chap. 3, environmental racism occurs when environmental decision-making becomes racialized to benefit one race over another. Sometimes, environmental racism can be overt, such as when an organization knowingly pollutes one community knowing it will be powerless to complain or litigate the pollution action. At other times, environmental racism can be systemic and be built within the implied structures of an organization or government. For example, a government that regularly zones polluting activities in one neighborhood and places community amenities in another neighborhood has systemic racist policies that are long-standing. Over time, this type of racism leads to distinct special differences that have broad implications for a range of things such as property values, access to fresh and healthy food, access to work, and income.

Environmental racism matters because the outcome of it leads to a range of disparities—particularly health disparities associated with environmental pollution. There are a range of health disparities that have emerged in recent years including asthma and obesity. Asthma is a pulmonary disease caused by a range of air quality issues—including outdoor and indoor air pollution. In the United States, cases of asthma vary with Black Americans three times more likely to die from asthma than White Americans. Black women, in particular, have the highest rate of death from asthma. Why does this difference exist? The issue is complicated, but one of the key issues is that Black Americans are exposed to higher levels of pollutants than White Americans. While there are distinct air quality rules that seek to protect the American population from health problems associated with air pollution, the evidence shows that the protection is not equal. Black citizens of the United States continue to be exposed to pollutants at rates disproportionate to White Americans.

The obesity epidemic throughout the world often has racial dimensions. To return again to the United States, a country where a great deal of work has been done on environmental racism, the highest levels of obesity are found within Black and Latinx populations. The Centers for Disease Control notes that while not all of the causes for the disparities are known, the differences are linked to social, economic, and environmental factors. One issue that has garnered considerable attention in recent years throughout the world is the issue of food deserts. Food deserts are places where there is very limited access to fresh and healthy food (Fig. 12.2). These landscapes are typified by a preponderance of inexpensive fast food and bodegas that sell fatty foods and sugary drinks.

The distribution of food deserts is mainly associated with poor, minority neighborhoods. The deserts can also exist in rural areas or small towns where grocery stores may be some distance away from human settlements. In urban areas, the issue of food deserts is often compounded by limited access to mass transit. Thus, even if grocery stores are relatively close, carless families may find it difficult to bring fresh and healthy food into the home.

It must be noted that environmental disparities can also have gender and age dimensions. Women, children, and the elderly are often in

Fig. 12.2 Not all people have access to fresh and healthy food such as is found in this market in Alexandria, Egypt

vulnerable positions in society and thus have some of the greater risks to disproportionate impacts from environmental racism. Thus, when considering how projects impact communities, special consideration should be given to these vulnerable populations.

Environmental Justice

Environmental justice seeks to shed light on environmental racism and work toward equitable solutions to the problems that caused the issue in the first place. After the murder of George Floyd in the United States in 2020 and after the murder of several indigenous environmental leaders in the Amazon rainforest in the last few years, we know that we live in a racist world. Bad things happen to minorities and people we define as "others" in our societies all over the planet. How do we work to solve the inequalities we all know exist?

Some governments have worked to institutionalize a process toward achieving environmental justice. For example, the US Environmental Protection Agency opened an Environmental Justice Office in 1992 and Presidential Order 12898 in 1994 required all federal agencies to consider the impacts of their efforts on environmental justice. Since its inception, the EPA's Environmental Justice Office sought to examine how the United States could better ensure that the benefits and burdens of environmental decision-making were shared among all communities. This ideal is clearly difficult, but progress has been made. One of the key tools has been the implementation of stakeholder engagement in decision-making.

Stakeholder engagement seeks to get input on all aspects of a project in consultation with the community impacted by the project. Thus, if a city wants to build a new waste to energy power plant in a particular area, it should seek input from everyone involved with the development of the project to make sure that all impacts are discussed, that the impacts are equitable, and that key decisions are made in consultation with the community. In true consultation with the community, the project can be modified in order to ensure that any impacts can be minimalized and to ensure that all impacts are clearly articulated prior to the project moving forward.

Over the years, the EPA has developed expertise in not only stakeholder engagement but in helping communities achieve their environmental justice goals. The office provides grants for community engagement activities and will provide expert advice on environmental justice issues. In addition, the office provides access to EJSCREEN, an online environmental justice mapping tool (EPA, 2020).

EJSCREEN uses 11 environmental indicators, 6 demographic indicators, and 11 environmental justice indexes in providing the public a mapping tool for assessing environmental justice throughout the United States. It is worth reviewing these indicators to get a better idea on the kind of variables that can be assessed when looking at spatial disparities in environmental justice.

The 11 environmental indicators are:

- National-Scale Air Toxics Assessment (NATA) air toxins cancer risk—focuses on lifetime cancer risk from inhaling air toxins
- NATA respiratory hazard index—ratio of exposure concentration to health-based reference concentration
- NATA diesel particulate matter—diesel particulate matter level in the air
- Particulate matter—particulate matter less than 2.5 micrometers
- Ozone—seasonal average of daily maximum eight-hour concentration in air
- Traffic proximity and volume—count of vehicles at major roads within 500 meters
- Lead paint indicator—percent of housing units built pre-1960
- Proximity to risk management plan sites—risk management plan sites are those that use extremely hazardous substances
- Proximity to hazardous waste facilities—count of hazardous waste facilities within 5 km including TSDF (treatment, storage, and disposal facilities) and LQG (large quantity generator) facilities
- Proximity to National Priorities List sites—nearness to Superfund sites
- Wastewater discharge indicator—modeled toxic concentrations at stream segments within 500 meters of wastewater discharge

The six demographic indicators are:

- Percent low income—measured at the block group (census indicator) level as population in household that is less than or equal to twice the federal poverty level
- Percent minority—percent of individuals in a block group that are other than white alone and/or list their ethnicity as Hispanic or Latino
- Less than high school education—percent of people of age 25 or older in a block group without a high school education
- Linguistic isolation—percent of people in a block group living in households with all members over 14 years of age who are non-English speakers
- Individuals under age of 5—percent of people in a block group under the age of 5
- Individuals over age 64—percent of people in a block group over the age of 64

The 11 environmental justice indicators are a combination of the environmental and demographic indicators and are:

- National Scale Air Toxics Assessment Air Toxics Cancer Risk
- National Scale Air Toxics Assessment Respiratory Hazard Index
- National Scale Air Toxics Assessment Diesel Particulate Matter
- Particulate Matter Less Than 2.5 Micrometers
- Ozone
- Lead Paint Indicator
- Traffic Proximity and Volume
- Proximity to Risk Management Plan Sites
- Proximity to Treatment Storage and Disposal Facilities
- Proximity to National Priorities List Sites
- Wastewater Discharge Indicator

What is interesting about this mapping tool is that it provides an open access tool for anyone in the world to conduct an assessment of environmental justice issues anywhere in the country. If you live in the United States, you can access the site here https://www.epa.gov/ejscreen and conduct a search for your community and examine all of the 11 environmental indicators, 6 demographic indicators, and 11 environmental justice indicators.

The United States' EPA tool, EJSCREEN, is a great example of how a national, state, or local government can assess environmental justice. However, as noted earlier in the chapter, there are many environmental justice issues that have their root in the action of international corporations that do not have distinct loyalty or regulatory responsibility to one particular nation. As a result, a number of standards have been developed to promote a range of sustainability initiatives and some of these will be reviewed in the next chapter. However, it is worth reviewing one of these, ISO 26000, since it focuses so heavily on corporate social responsibility (ISO, 2020).

ISO, or the International Organization for Standards, began in the 1940s to promote international standardization of a variety of measurements. Early on, the group focused on basic standardization of a range of scientific units such as length or weight. However, with time, the group developed a range of measurement standards that are widely used in manufacturing and shipping. For example, the ISO was largely responsible for the development of shipping container standards that promote ease of transport of containers from country to country on ship, rail, and truck.

In 2010, the ISO published ISO 26000, which included a set of guidelines around corporate responsibility. This document evolved, in part, due to the wide impacts of globalization. Citizens all over the world were concerned over many business practices that were emerging that exploited citizens in developing countries for economic gain in developed nations.

ISO 26000 is built around the idea that social responsibility should be a key ethical dimension of international corporate behavior. The standards are built around seven principles of social responsibility:

- Accountability: organizations must be accountable for their actions
- Transparency: decision-making and operations must be transparent
- Ethical behavior: organizations should abide by clear ethical standards
- Respect for stakeholder interests: communities that are impacted by an organizational activity should have their interests taken into consideration

- Respect for the rule of law: organizations should respect the laws of the nation, state, or community where they have any impact
- Respect for international norms of behavior
- Respect for human rights

ISO 26000 also follows six interrelated social responsibility subjects:

- Human rights
- Labor practices
- The environment
- Fair operating practices
- Consumer issues
- Community involvement and development

Specific goals can be set by organizations using ISO 26000 guidance. For example, a company can focus on child labor practice or on enhancing the circular economy to avoid waste issues. It is important to note that while most companies who work within the global marketplace utilize ISO standards for things such as packaging or weights and measures, they do not all embrace the social responsibility tenets of ISO 26000. Nevertheless, the ISO organization does provide a framework by which a range of stakeholders, particularly host countries and communities, can evaluate the nature of corporate responsibility toward environmental justice.

Ways You Can Advance Environmental Justice

Environmental justice is one of the hardest issues to address within the realm of sustainability, but there are concrete steps that you can take. These are listed in Table 12.2. At the most basic level, you can get to know the key environmental justice issues in your community and around the world. In addition, you can find ways to celebrate Human Rights Day each December 10th. To deepen your commitment to environmental justice, you can examine how you spend your money to make sure that you only support companies that have demonstrated ethical

Table 12.2 Simple, hard/expensive, and innovative/life-changing approaches to advancing environmental justice

	Ways to advance environmental justice
Simple	Understand environmental justice issues in your community
	Understand key environmental justice issues around the world
	Find a way to recognize and celebrate Human Rights Day on December 10 of each year
Hard/expensive	Evaluate how you spend funds to ensure that you only work with companies that support environmental justice and social justice
	Support organizations that work on environmental justice policies
Innovative/life-changing	Work with local groups to advance environmental justice in your community
	Advocate internationally for environmental justice through volunteer work

considerations around their interactions in the communities where they do work. This could be through the embrace of ISO 26000 or some other benchmarking system. In addition, you could support organizations that work on trying to advance environmental justice policies and laws in your community or around the world.

To further advance your support for environmental justice you can get involved in your community with environmental justice organizations to support the actions or activities that have meaning to local community members. You can also get involved internationally through volunteer work and advocacy. Environmental justice, environmental racism, and human rights require constant vigilance and some of the more important things that we can do is to shed light on inequity and to advocate for those harmed by bad practices and unethical behavior and decision-making.

References

EPA. (2020). EJSCREEN: Environmental justice screening and mapping tool. Retrieved December 5, 2020, from https://www.epa.gov/ejscreen

ISO. (2020). ISO 26000—Social responsibility. Retrieved December 5, 2020, from https://www.iso.org/iso-26000-social-responsibility.html

Markowitz, G., & Rosner, D. (2013). *Deceit and denial: The deadly politics of industrial pollution.* University of California Press.

Mielke, H. W., & Zahran, S. (2012). The urban rise and fall of air lead (Pb) and the latent surge and retreat of societal violence. *Environment International, 43*, 48–55.

Newsome, T., Aranguren, F., & Brinkmann, R. (1997). Lead contamination adjacent to roadways in Trujillo, Venezuela. *Professional Geographer, 49*, 331–341.

Riva, M. A., Lafranconi, A., D'Orso, M. I., & Cesana, G. (2012). Lead poisoning: Historical aspects of a paradigmatic "Occupational Environmental Disease". *Safety and Health at Work, 3*, 12–16.

United Nations. (2020). Universal Declaration of Human Rights. Retrieved December 5, 2020, from https://www.un.org/en/universal-declaration-human-rights/

13

Green Your Economy

Introduction

In our modern global capitalist society, we have been moving away from a less is more ideal as espoused by modernist architect Ludwig Mies van der Rohe. Now, more is more. Even in self-proclaimed socialist societies like China and Venezuela, excess is the norm and the elite strive for more stuff and more glitz. Capitalism is in overdrive all over the world. We have access to a range of cheap goods and services and spend as if there is no end to the party. The unfortunate outcome of all of this is that there are significant planetary impacts from all of this economic activity.

The negative planetary impacts of global capitalism have long been discussed. Perhaps the most influential book on the topic was written by E. F. Schumacher in 1973 and is called *Small Is Beautiful: A Study of Economics as if People Mattered* (Schumacher, 1973). One of the key concepts of the book is *enoughness*—when do we decide when we have enough stuff. This idea, influenced in part by Buddhism, questions the twentieth-century drive to consume more and more goods and resources. Indeed, Schumacher highlights that up until the 1970s natural resources were treated as if there were no end to them and that pollution was not a significant limiting factor. Of course, the energy crisis in the 1970s demonstrated that there were indeed limits to natural resources that could be provided by the planet. In addition, pollution problems were rampant in the 1970s and it was clear to any cognizant observer that the planet was unable to process the volumes of waste that we were creating. Schumacher suggested that we needed to question our use of technology and our drive toward consumption. Similar books around this period, for example, *Diet for a Small Planet* (1971) by Frances Moore Lappé, questioned a range of twentieth-century activities and urged social change around economic decision-making (Lappé, 1971).

Since the 1970s, of course, globalism has accelerated. The problems highlighted by Schumacher have not disappeared. Indeed, Naomi Klein's 2014 book, *This Changes Everything: Capitalism vs. the Climate*, demonstrates how global capitalism has shaped the key environmental challenges we are facing around climate change (Klein, 2014). She points out that neoliberal policies that brought about our modern globalism positioned us on the brink of planetary disaster. The challenge, of course, is that global neoliberal policies are powerful forces in our world right now and most countries are deeply integrated into this modern economic system and it is particularly difficult to change the status quo when so much money is in play.

However, there are important developments that have changed the way we think about and interact with the modern neoliberal economy from a sustainability lens. This chapter highlights three major areas: green businesses, green economic development, and green investing and finance. The chapter concludes with actions that you can take to advance the green economy and make a difference on the planet.

Green Businesses

The idea of green business is complicated because there are so many different approaches to what people consider a green business activity. At the most basic level, a green business is a company that seeks to have a positive impact on the planet or one that seeks to have a smaller impact than is traditional within their economic sector. Obvious examples of

Fig. 13.1 This small organic farm is an example of a green business

green businesses include organic farms, solar energy companies, or an ecological restoration firm (Fig. 13.1). However, there are many companies that are making clear measurable changes to their business models that could be considered green businesses even though they may not be perceived as a green business by the consumer.

An example of this type of organization is Walmart. Walmart has gotten plenty of bad marks over the years for driving out small local retailers around the planet. In addition, because they provide cheap, abundant consumer goods, they are considered a major driver of consumerism. Plus, many retailers like Walmart have also been responsible for well-documented cases of poor labor practices in the developing world. So how could they be considered a green business?

In fact, in 2006, Walmart was one of the first major retailers to embrace sustainability within their corporate model. At first, many of their sustainability initiatives focused on the environmental E of the three E's.

They worked to reduce packaging, shipping costs, and energy use within their supply chain. They also sought to advance green energy initiatives and reduce energy use within their stores. They developed company-wide sustainability initiatives lead by a team of corporate sustainability experts. Indeed, they were one of the first companies in the world to hire a broad comprehensive sustainability team.

Since the early days of Walmart's efforts, they have deepened their approach to sustainability by developing a series of benchmarking tools to evaluate their supply chain that includes a range of indicators within the three E's. While not everyone will embrace the nature of Walmart's business model, there is no doubt that they can be recognized as one of the first global companies to infuse sustainability within the retail world. Since then, many companies have found ways to infuse sustainability across the three E's of environment, equity, and economy, or in the business lingo, people, planet, and profits.

What is important within all of these corporate efforts is that they are transparent and measurable. The public has a deep distrust of greenwashing, or the use of "feels" of sustainability in inauthentic ways to market a business. One of the most notorious examples of greenwashing is the use of beautiful photographs of natural spaces in advertisements for fossil fuels. Of course, energy companies are not going to show the reality of oil spills or climate change when they are trying to sell us all energy. Instead, they try to make us feel good about their products by linking their product to the very things that are threatened by the use of their product.

Many major global companies from Walmart to Unilever have significant sustainability initiatives. However, green businesses more typically refer to organizations that market or sell green products like organic food or that offer services, like yoga classes, that are linked to simpler lifestyles. While the term, green business, is ill defined, we more or less know them when we see them.

Over the years, some communities have developed special support or incentives for green businesses. For example, around the world, green chambers of commerce have cropped up to support green businesses and entrepreneurs advance green business agendas. In the United States, the US Green Chamber of Commerce started in 2011 and the European Chambers of Commerce listed the low-carbon and circular economy as one of their top priorities. Organizations like the Green Business Bureau help to educate business leaders about the benefits of embracing green approaches around their business practices and also offer benchmarking opportunities that help to guide businesses toward communicating their efforts to their stakeholders.

The key point of all of the green business approaches is that nearly every business can make some improvements to help drive the sustainability agenda forward. Any business activity can look deeply at their activities to try to find ways to make interventions into their daily activities to demonstrably engage with the three E's of sustainability. Plus, businesses that want to move toward more sustainable practices can often take advantage of green economic development initiatives as discussed in the next section.

Green Economic Development

Green economic development occurs when leaders in a particular region focus on advancing a broader economic development policy that prioritizes sustainability and environmental protection. It is in direct contrast to green business development in that green business focuses on a single entity or sector. Green economic development is much more holistic and engages with all sectors of the economy.

Economic development tends to be driven by national, state, regional, or local governments and business leaders. Perhaps one of the most famous economic development initiatives ever undertaken in the world was the Marshall Plan which invested billions of dollars toward the economic recovery of Western Europe after World War II. The plan sought to bring back productive businesses in the region by modernizing industries, cutting barriers to trade, and by working with local assets to enhance job growth. There is no doubt that Europe was on the way toward recovery without the Marshall Plan, but the infusion of dollars helped accelerate economic growth.

Throughout the twentieth century and into our modern century, countries around the world have tried to plan their economies. Some developed five-year plans around industrialization or modernization of rural areas, while others developed plans around particular industrial sectors in key regions. Some nations try to force economic development plans on unwilling populations while others use financial incentives to advance economic development goals. In post-World War II Romania, for example, Nicolae Ceauçescu, that country's leader from 1967 through 1989, implemented a variety of industrialization schemes that caused great disruption throughout the country. Indeed, poor centralized economic planning led, in part, to the Romanian revolution that caused his downfall and execution by firing squad in 1989. Clearly, if economic development is done poorly, and without the support of the public, bad things can happen.

The idea of infusing green initiatives within economic development has gained in significance in the last 30 years. However, it has become relatively common within the last ten years. Essentially, green economic development focuses on building an economy that advances the sustainability goals of a region. These initiatives can vary greatly depending on the existing economy and geographic setting and natural assets of an area.

Greening Existing Economies

Some parts of the world focus on finding ways to advance sustainability within green economic development strategies. In Milan, Italy, for example, the Fondazione Cariplo has been funding grants to advance research on how to improve the circular economy of Milan's intensive industrial sector. The research funded by the foundation is helping industries and government better manage their resources while helping to create new jobs across northern Italy.

All over the world, organized and planned investments into green initiatives help drive green economic development strategies. In Michigan, for example, the auto industry has taken advantage of federal and state grants to enhance the use of green energy in their facilities and to advance energy-saving technologies in vehicles.

Focusing on Natural Assets. Another important green economic development strategy is to focus on natural assets of a region to drive regional economic goals. There are many examples of regions that have taken advantage of this strategy. The windy US state of Texas, for example, has invested heavily in wind energy to help the state become the leading producer of wind power in the United States. Texas' unique geographic setting helped move some of the wind projects forward with investment help from the state and the US government.

The island nation of the Bahamas also seeks to advance a green economic agenda by encouraging tourism on some islands while also protecting and preserving natural assets in others by limiting development and access. Because the Bahamas, and other nations with great natural tourist attractions, are so effective at protecting their environments, the beauty of the natural areas helps to enhance and increase tourism and thus economic growth.

In Long Island, New York, the local regional economic development council has focused attention on protecting agricultural and fisheries resources by investing in small farms, organic operations, shell-fishing operations, and protection of natural areas. By integrating small-scale agriculture within their plan, they are enhancing the Long Island food systems and creating networks across the island that help to preserve rural landscapes that are constantly under threat from the massive growth of the New York City region.

How Is Green Economic Development Done? Most regions of the world have some type of economic planning organization. At the most basic level, there may be local chambers of commerce that try to do some type of economic development planning. However, there are often regional, state, or national organizations that set goals and priorities for an area. Goals may be in the form of job numbers, job types, industrial sectors, educational priorities, or societal improvements. When sustainability is infused within goal-setting initiatives, new priorities emerge. For example, an economic development group may seek to have goals like: develop green energy sources, enhance land preservation to promote tourism, create a green business rating system, reduce the production of waste in the area, advance career training and education for a green economy, and so on.

Goals are often set by economic development councils that are made up of a variety of stakeholders that include elected officials, business leaders, non-profit leaders, educational leaders, labor experts, and other community members. Funds are obtained through national, state, or local sources for investment in the goals designated by the community. Once the goals are set, the region can invest available funds toward the goals. For example, a group may provide grants for companies seeking to develop green energy projects or may give matching funds to local communities to purchase land for parks or environmental preservation. Or, funds may be given to universities or community colleges to start new programs on things like green energy or organic food based on anticipated green job growth in the area. The efforts of the councils are usually reviewed and evaluated each year to ascertain if the investments have led to success.

There are critiques of the green economic development approach that largely emerge from broader critiques of capitalism. Any economic growth can lead to enhanced consumption of natural resources. Plus, the green part of any region's economy tends to be relatively small and most economic development plans are not comprehensively "green" in their approach. Thus, green economic development planning can lead to local greenwashing and green boosterism without significant outcomes. Having one organic farm supported via a region's economic development dollars that is located in a landscape full of conventional farming does not make the entire region a green region. Unfortunately, some places tend to push forward small, green projects like organic farms, small renewable energy projects, or local backyard chicken ordinances, without broader regional infusion of sustainability across the region's economic sectors.

Regardless, of the critiques, green economic development can be a powerful tool for communities that seek to transform themselves. Often, green economic development takes on a technological or modern vibe that helps to drive community investment and draw people to join in on green projects from outside the region. Thus, some places have used green economic development strategies to help reduce their local brain drain. Many parts of the world are seeing population declines and are trying to find ways to draw new residents to their community. A clear commitment to green economic development is one community revitalization strategy that can help.

Green Investing and Finance

The finance world has changed a great deal in the last few decades. Financial markets around the world have increased greatly and the world is heavily interconnected within a global financial market. In the past, financial markets tended to be localized. While some financial centers like London and New York were somewhat global in scale, they tended to focus on national or regional financial initiatives (Fig. 13.2). The growth of the Internet and other forms of communication such as Zoom opened the borders of the financial world. Financial decisions that used to be made in the board rooms in the world's financial capitals can now be made anywhere. Companies can share public information on websites and also store financial information on cloud-based services so that stakeholders, regardless of time zone, can engage and interact with financial leaders from anywhere. Thus, decisions made about expanding

Fig. 13.2 New York City is home to the United Nations and many Fortune 500 companies like BlackRock that are making the world's economy greener

operations of an Austrian company in Australia can be made from the Bahamas. This ease of information helps drive global finance in new and innovative ways—some for the good, and some for the bad.

One of the most interesting aspects of modern finance is that the average person can find out much more information about companies using the Internet than they were able to find in the past. Most firms have some type of basic information available on the Internet that helps people understand the impacts of financial decisions on their community, country, or planet. If, for instance, you are an investor interested in investing in companies that focus on issues of green energy, you can easily find them and find ways to invest in them. Because of the increasing interest in sustainability and the environment, many companies highlight their green initiatives in order to make them more attractive to investors who prefer to put their money in these types of investments. A simple Web search on any Fortune 500 company's website for the term, *environment*, or *sustainability*, will undoubtedly provide a range of information about the green initiatives of the organization. However, it is up to the consumer to assess whether or not the information is accurate or if it is a form of greenwashing.

Buying individual stocks can be financially risky. As a result, a number of investment firms have developed a range of products that cluster a group of funds together. Managers of the funds take the money that is entrusted to them and invest it in a range of businesses. Perhaps one of the most famous funds is one that is tied to the Fortune 500 companies. These are the strongest firms in the world and thus funds that are invested in them are likely to reap strong benefits for investors. If one company has a bad year, there are 499 other companies to make up the difference. Thus, by pooling funds to invest in a group of businesses, there is usually less risk to the investor. These funds tend to move broadly up and down with the economy and have less volatility than investments in single businesses.

Over the years, these investment firms have developed innovative strategies to meet the needs of their clients by creating investment clusters based on the affinity of their clients for particular market sectors or interests. For example, some clients may wish to invest in a particular sector of the economy like services or manufacturing. As a result, there are funds that bundle purchases of stocks in several similar companies to meet the clients' needs and to protect the investment by grouping stock purchases together.

One cluster of investments that has gained traction in recent years is socially responsible investing (SRI). Investment firms bundle stock purchases of socially responsible companies together to provide investment options for people who want their money to work for the greater good of society. Of course, the biggest challenge is in defining what is meant by a socially responsible company. Larger investment firms do a tremendous amount of research on the stocks they purchase and report back the impacts of the SRI funds to the owners of the funds. For example, they may purchase stock in a company that provides loans to low-income households to promote homeownership and report back the number of loans or homes provided through the process. Or, they may invest in companies that commit to paying good wages and providing excellent benefits to employees. Some have criticized this form of investing as not being clear enough on the reporting standards and that they provide a "feel good" approach to globalization and modern neoliberal economic systems. Regardless, this type of investment has helped drive change in business practices to attract ethical investors.

One type of socially responsible investment is green investing/green finance. This form of SRI involves the investment of dollars specifically on sustainability initiatives. For example, such a fund could invest in green energy, organic agriculture, and sustainable forestry. Those who purchased shares of the funds would know that their dollars are going to support their own personal interests in sustainability. Most major investment houses around the world have some SRI products focused on sustainability and the environment.

Along with the growth of SRI products, there has also been a cultural shift to encourage divestment of fossil fuels due to the impact of fossil fuels on global climate change. Inspired by the overall success of the divestment from the South Africa campaign a generation ago, today's investment activists are encouraging large investment holders, particularly colleges and universities, to divest from fossil fuel companies and other businesses that in some way do damage to the environment.

In the last few years dozens of colleges and universities are moving their hefty endowment dollars away from fossil fuels and toward socially responsible investment options. For example, the University of California, with their massive 83 billion dollar endowment, moved away from fossil fuel investment in 2019. But colleges and universities are not the only organizations making change. Many religious organizations, including the Vatican Bank, have divested from fossil fuels as have many government pension fund operators. Trillions of dollars have moved away from the fossil fuel industry in this social movement. Green investing and finance is not just about where to put your money for the greater good. It is also about making sure that your money is not used to harm the planet in some way.

As noted, most major investment houses offer some form of socially responsible investment. However, in 2020, the largest investment firm in the world, BlackRock, changed the game by stating that they would no longer invest in organizations that did damage to the planet in some way. In a letter to shareholders, the CEO of BlackRock, Larry Fink, outlined his vision as to how the company is going to react to the problems brought forward by climate change. He highlights how climate change is a risk to the status quo:

> *Climate change has become a defining factor in companies' long-term prospects. Last September, when millions of people took to the streets to demand action on climate change, many of them emphasized the significant and lasting impact that it will have on economic growth and prosperity – a risk that markets to date have been slower to reflect. But awareness is rapidly changing, and I believe we are on the edge of a fundamental reshaping of finance.*
>
> *The evidence on climate risk is compelling investors to reassess core assumptions about modern finance. Research from a wide range of organizations – including the UN's Intergovernmental Panel on Climate Change, the BlackRock Investment Institute, and many others, including new studies from McKinsey on the socioeconomic implications of physical climate risk – is deepening our understanding of how climate risk will impact both our physical world and the global system that finances economic growth.* (BlackRock, 2020)

The letter marked a fundamental shift in BlackRock's investment approaches. Indeed, the letter signaled to the world that the investment world was changing. They needed to drive change beyond offering socially responsible investment options. They needed to infuse sustainability across all areas of the investment and finance industry. The pressing issue of climate change is too important of a risk to allow the investment world to ignore.

While BlackRock provides a positive example as to how the investment world is changing, a recent scandal involving Goldman Sachs shows how the lure of money can take good intentions down a dark path. Goldman Sachs Inc. is one of the most influential global financial institutions in the world. Founded in 1869, it has specialized in a range of big financial projects—particularly the management of national government funds. One of the things it specializes in is putting together large funds for investment in national priority projects.

In 2009, the prime minister of Malaysia, Najib Razak, put together a fund called 1MDB to invest in a number of development projects that included major components of sustainable development for the country. Thus, the fund could have been seen as an SRI type of fund. The fund raised billions of dollars of investment dollars. However, much of the money was siphoned off by individuals working for Goldman Sachs to accounts held by Razak and others. The corruption started to come to light in 2015 and several individuals, including two high-ranking Goldman Sachs employees, were charged with their part of the scheme. Tim Leissner, Chairman of the Southeast Asia branch, pled guilty to money laundering and forfeited millions of dollars as part of his plea. It is one of the most notorious financial scandals of our generation.

The 1MDB scandal highlights how easily major institutions can be corrupted. Leissner claims that leaders at Goldman Sachs knew of his dealings and overlooked them since the fund was making Goldman so much money. The real losers in this scheme, however, are the Malaysian people. They trusted that their government was working on sustainable development projects through this well-publicized fund. Instead, the money was diverted to individuals who used it to support lifestyles that few in Malaysia could hope to attain. It caused many around the world to lose faith in global finance and in global funds used to support sustainable development.

There is no doubt that there are bad actors in the financial world who advance socially responsible investing for their own corrupt purposes. However, financial institutions, if they are ethically managed with appropriate oversight from boards and regulatory agencies, can be a force for good. They can help move economies in particular directions for the betterment of all of us.

What You Can Do to Advance a Green Economy

How we spend our money is one way to advance a green economy. However, we interact with the economy through our place of work and through our interactions with the broader community in which we live. Thus, we should take a holistic approach to consider how we can actively help to transform our society so that it embraces more of the ideals of the green economy. Table 13.1 provides some guides for some things that we can do to help move the green economy forward.

Table 13.1 Simple, hard/expensive, and innovative/life-changing approaches to advancing a green economy

	Ways to advance a green economy
Simple	Support green businesses that sell certified products such as those certified by The Rainforest Alliance
	Invest in socially responsible funds
Hard/expensive	Evaluate how you spend all of your money and evaluate how you can spend your money on green businesses
	Evaluate where your investment funds go and divert funds to green socially responsible funds
	Join a local chamber of commerce or other business group to help support green economic development
Innovative/ life-changing	Start your own green business
	Transform your place of work into one that takes on a broader green mission
	Actively work to promote green economic development in your region

At the most basic level, we could all support those businesses that support green initiatives. Some of these may include green-certified products like organic food or tropical products produced under the Rainforest Alliance certification program. We can also support small businesses that focus on local green products or in some way give back to the community through their actions. We can also find ways to invest in socially responsible funds. There are a range of products on the market that help investors find the appropriate fund for their interests.

To deepen your commitment to the green economy, you can take a deeper dive into how you spend all of your money. You can look at how you spend all of your money and find ways to transform your choices so that you support more green businesses. For example, if you have a mortgage, do you have it with a bank that supports the green economy? Each dollar we spend either can be used to support a green business or can go into the broader economy that supports the status quo. By being more intentional about how you spend every dime you help to drive change. You can also deepen your understanding of your investments to make sure that they go into green socially responsible funds. Finally, you can join your local chamber of commerce or other business group to learn more about the economy in your region and to influence it to advance green economic development.

If you want to take on innovative or life-changing approaches to the green economy, you can start your own green business. There is more and more support for green businesses than ever before. People are drawn to green products and services because they know that they are better for the planet. If you are unable to start your own business, you can work to transform your place of work, worship, or education. Find ways to make differences and influence decision makers. Share ideas on how to green different units of your organization. For example, most organizations buy lots of stuff. You can share how procurement could be done differently. You could also help drive your organization into purchasing green energy credits or into divesting from fossil fuels. Finally, you could get actively involved in regional economic development efforts managed by state or national organizations to push them into green economic development and away from activities that will do environmental or social harm to your community.

References

BlackRock. (2020). A fundamental reshaping of finance. Retrieved December 5, 2020, from https://www.blackrock.com/corporate/investor-relations/larry-fink-ceo-letter

Klein, N. (2014). *This changes everything: Capitalism vs. the climate.* Simon & Schuster.

Lappé, F. M. (1971). *Diet for a small planet.* Ballantine.

Schumacher, E. F. (1973). *Small is beautiful: A study of economics as if people mattered.* Blond & Briggs.

14

Sustainable Travel and Leisure

Introduction

Due to labor laws regarding vacation time, many of us around the world are finding more time on our hands for travel and leisure. This is a wonderful development in human history. The industrial revolution created very challenging labor conditions for workers and it took decades for managers and business owners to develop the appreciation for worker time off. Giving employees vacation time creates a healthier and happier workforce. Some countries require up to a month off a year, while others provide one to two weeks. Some countries do not specify vacation time, but have business standards that range from two to four weeks of time off for their labor force. Regardless of the amount of time off, most of us have to decide how we spend our time away from work.

The decisions we make around our time off do have impacts on the environment. In the last few decades the options for tourism have grown as the world has become more connected. Plus, cable travel channels feature ideal vacations that people can take to exotic locations. No longer is it rare for people to travel great distances for relatively short time periods for vacations. We can jump on an airplane to Kyoto, Tahiti, Luxor,

Bilbao, Buenos Aires, or New Orleans from almost anywhere in the world. A vast network of hotels and local tour operators are available online so you can easily plan complex trips that would have been difficult to organize a generation earlier. Plus, the world, for the most part, is relatively safe for tourism. Even in countries where there is civil unrest, such as in Egypt or the United States, tourist sites are kept safe.

For some countries and regions, tourism has been accentuated as a form of economic development. The nation of the Bahamas, for example, recognizes that their beautiful islands, beaches, and waterways are major tourist attractions for people who need a getaway from cold climates. As a result, they have built significant infrastructure around tourism. Most of us have experienced some of the places around the world that have built grand tourist infrastructure. Places like the Bahamas, Las Vegas, Orlando, Luxor, and Rome are some big examples, but there are smaller ones that have become regional centers such as Door County, Wisconsin, and Margarita Island in Venezuela. These smaller regional centers often are built around some type of natural asset.

Regardless of size, these tourist centers have impacts on the environment and on local populations and there is growing interest in how we can reduce the impact of our leisure choices on the host site. Our impacts take many forms. I lived in Florida for many years and I often saw tourists doing things that were bad for the environment. They would disrupt sea turtles or manatees in their habitat, they would drive boats in protected areas, and they would leave trash on the beach or along wetland hiking trails. Local Floridians both loved and hated the tourists. They loved them because they brought money to the state, but they hated them because of the increased traffic and the environmental problems they created.

In my own travel, I have felt the wrath of local people around outsiders. I was in Riga, Latvia, for business where I saw large groups of men who traveled from Great Britain for stag parties due to the low cost of travel to this Baltic country. The groups of men were publicly drunk at inappropriate hours and were disrespectful of local Latvians on the downtown streets. Shortly after my visit, the mayor of Riga cracked down on some of this behavior. On a visit to Amsterdam, I saw a young couple throw used food containers into one of the city's famous canals. A local resident yelled at the hapless couple who tried unsuccessfully to reach the garbage before it floated away.

The point is that we all have some type of impact when we travel and how we travel and how we behave when we travel matters. This chapter reviews several aspects related to sustainability and travel. It explores a range of ideas starting with a case for staying local. Following this, the chapter focuses on green travel, green cruises, green lodging, ecotourism, and volunteer tourism. The chapter concludes with ways that you can make greener transportation choices.

The Case for Staying Local

Our modern world tends to fetishize exotic travel over local experiences. In many ways, exotic travel has become a type of luxury good that we want to have. We have created bucket lists (lists of things we want to do before we die, or "kick the bucket") with places we want to see before we die or things we want to experience. We want to go to Paris, see the pyramids, skydive, or climb a mountain. All of these experiences are designed to take us away from the ordinary. For some, these experiences are thrilling and well worth the time and effort to make them happen. For others, the experiences are disappointing and seem like just another day—just a different view. We are who we are regardless of where we are and the experience of exotic tourism is sometimes not all that rewarding.

Many of us who travel to exotic settings rarely take the time to explore our own local region. Every place is interesting in some way and worthy of deeper understanding. I currently live in DeKalb, Illinois, a small college town in north-central Illinois. The region is surrounded by corn and soybean fields that go on for miles. What travel and tourism options are available here? In reality, the region has miles and miles of amazing hiking and biking trails and Chicago with its great museum and theater venues is just an hour and a half away. Even modest Gary, Indiana, which is part of the Chicagoland region, has a vibrant art scene with many public murals and historic sites.

Taking time to get to know an area opens up a range of possibilities for a deeper understanding of yourself and your community. There are local authors who have taken the time to get to know their region. For example, the novelist Tom Dorsey, who has written over 23 novels about Florida, provides local readers a deeper understanding of the craziness and beauty of the Sunshine State. Historical societies and museums also help to get a better sense of place. Local social clubs and chambers of commerce also help provide local information about things to do and places to see.

In my community, there are local apple orchards that provide hay rides in the fall and opportunities to pick apples and taste apple cider. During the COVID-19 lockdown, self-guided barn quilt tours were organized. Barn quilts are geometric designs painted on barns which are a local expression of creativity across some areas of rural America. In addition, the rural communities in my area highlight restaurants, bakeries, breweries, confectionaries, and other attractions that tourists might enjoy. If I get tired of that, I can always go online and take a virtual tour of some exotic locale.

We can travel thousands of miles and not find the joyful things that are available within a short radius of our home. However, by staying local, you can help your local economy, gain a better sense of place, and also lower the ecological footprint of your travel choices. As we will see in the next section, there are green options for travel, but they still have a larger footprint than if you stayed local.

Green Travel

Travel is how we get to our tourist or work destination. Green travel is travel that lowers our ecological footprint in some way. Some people use green travel as an overarching term to mean any aspect of travel and tourism that involves sustainability in some way. However, as noted in the introduction of this chapter, there are very specific terms that can be used for sustainable approaches to travel and tourism and I opt to use the term green travel to refer only to the transportation portion of any trip.

There are many options for lowering the ecological footprint of any trip. As was discussed before, airplanes by far have the greatest carbon footprint of any transportation type. This is part of the reason that many green travel advocates have focused so much on the local and not the exotic or distant. There are a number of initiatives underway to green air travel. For example, an airplane called the Solar Impulse, which was entirely powered by solar energy made an around the world trip in 505 days in 2015 and 2016 (Solar Impulse Foundation, 2020). The travel was slow due to the need for recharging and due to the needed weather conditions for the light vehicle. Nevertheless, the trip provided a proof of concept that air flight can be carbon neutral.

Since the Solar Impulse made its historic flight, many airplane manufacturers have developed a range of green initiatives for airplanes for large aircraft as well as for smaller private planes. While solar airplanes are not the norm at this point, the proof that they can work is a key step in developing the technology to make them more common. At the moment, they require considerable investment due to the high cost of the technology. In addition, they are very light and can only fly under particular conditions.

Car travel can also contribute significantly to greenhouse gas emissions. While there are electric cars on the market, most typically do not go but a few tens of miles before they need recharging. The Model S Tesla can go up to 400 miles on a single charge, but these cars are not widely used at this time. Plus, it must be noted that the massive road infrastructure we have around the world is particularly damaging to the environment. There is no doubt that the future of cars and trucks is electric and that the world will improve technology to enhance the range of vehicles and improve the infrastructure for charging stations. Yet at the moment, tourists do not have many good options for long-distance vehicular travel to meet their tourism needs. Electric vehicles are fine for local travel but are not great options for long-distance vacations. Of course, you can always purchase carbon credits for any car or air travel.

Buses are another option for travel. Most countries have a range of long-distance bus options available for getting people from place to place. Some companies offer regular runs between nearby cities such as shuttles between Washington, D.C., and New York City, while others may run fully across the country. For example, Greyhound Bus offers a trip from New York City to Los Angeles that takes about two and a half days for around $210—far cheaper than driving yourself or flying. Greyhound and other bus companies often help to provide transportation access for small towns and cities that are not serviced by airports.

A number of tour companies offer group tours where participants in a tour meet at a particular city or hotel and then ride together on buses to see the sites. These experiences often do require some air, train, or car travel to get to the site where the bus departs, but overall creates a far less carbon footprint than if the people on the bus were traveling individually in cars.

Train travel is growing in popularity in many areas of the world largely due to the improvement in technology to enhance travel time. Bullet trains can go over 200 miles per hour, making them on par with air travel in some cases considering the time it takes to pass security and navigate through airport terminals. Because of the heavy infrastructure of train tracks, train travel is mainly available between large cities. Most train travel these days is powered by electrical energy which is often, in part, derived from green energy sources.

Green Cruising and Boating

Cruising as a form of tourism has really taken off in the last few decades (Fig. 14.1). In 2019, more than 27 million passengers took some type of cruise. This is up from under four million in 1990. The growth in cruising is closely tied to the rise in globalism and subsequent global connections. Tourism boomed during the last 30 years and cruising was one of the sectors of tourism that saw tremendous growth. As a result of this growth, cruise terminals have expanded as have the deleterious impacts of cruising on the environment. There are regular documented cases of pollution and litter from cruise ships and even environmental impacts when cruise ships run aground. In addition, cruises tend to have limited "real" cultural experiences for tourists at the ports that they visit since they are docked for relatively limited time periods.

Some areas, such as Nassau, Bahamas; Cozumel, Mexico; or Sanya, China, have well-developed port facilities and a range of cultural experiences for tourists. For example, in Cozumel, tourists can visit Mayan ruins or explore natural wonders of the Yucatan. Many of the natural

Fig. 14.1 Cruising has gained in popularity in recent years, but some cruising is greener than others. This small cruise ship on the Nile, called a dahabieh, uses wind for part of its power

areas, such as the reefs off of Mexico and Belize, have been damaged by the increase of tourism. In addition, due to the highly controlled nature of the port experiences, few local people benefit from the tourist dollars that enter the areas.

There are some green initiatives around cruising. For example, the Friends of the Earth rates cruise lines based on four criteria. The first one focuses on sewage treatment. In the past, cruise ships released raw sewage directly into the ocean. Now, there are excellent treatment facilities available, but not all cruise lines utilize the technologies. The second criterion focuses on air pollution reduction. Cruise ships burn some form of oil for running their engines. As a result, there are major pollution problems. These problems can be harmful for local populations if a cruise ship continues to run its power at dock using oil. Port electrical plug-ins are available in many areas, but not all cruise lines take advantage of this technology. The third criterion focuses on water quality compliance associated with pollution standards in local areas. Finally, the last criterion deals with whether or not the cruise companies are compliant in providing transparent responses to the Friends of the Earth's request for information on the previous three criteria.

Of course, when people think of cruising, they think of the large cruise ships that hold hundreds of people and that have a limited set of ports of call like Miami, Shanghai, or Barcelona. However, there are a range of other cruising options that offer greener options. They are more expensive, but they provide options for people who are serious about lowering their ecological footprint. Tall ships that utilize only wind power are a great example of how we are returning to simpler technologies in sailing. The Atlantic crossing by Greta Thunberg in a racing yacht showed how we can take advantage of wind power to advance international travel without the use of fossil fuels.

Boating or yachting is another way that people enjoy leisure time on the water. The terms boat and yacht are sometimes used for the same type of craft. However, a boat is generally considered smaller than a yacht. Indeed, yachts are often used as permanent or semi-permanent homes for some families due to their large size. Regardless, interest in recreational boating and yachting has grown tremendously in the last decade. There are many different types of boats and yachts. For example, small recreational fishing boats are obviously used for recreational fishing and speed boats are used for water skiing or pleasure boating. There are, of course, sailboats, pontoon boats, catamarans, and large yachts among the range of watercraft on the market.

The environmental impact of recreational boating has been well documented. Boaters, along with marinas that support them, are responsible for a range of pollution problems including oil and gasoline spills, sewage discharges, and litter. Recreational fishermen and women are responsible for vexing fishing line litter that can entangle birds and aquatic animals. They can also overfish and do damage to ecosystems. Uneducated boaters who take their craft into inappropriate areas are responsible for ecosystem damage from motors. There are many well-documented examples of boats doing damage to coral reefs, sea grass beds, and animals themselves such as the Florida manatee.

A range of greener choices are available to boaters such as canoeing, kayaking, and white water rafting. There are many tour companies that offer tours and guided paddles through some of the most beautiful waterways in the world. In addition, many people own their own kayaks, canoes, or paddleboards and take day trips in their own area to enjoy nature in a gentle, low-impact way.

Green Lodging

One of the most impactful parts of any tourist's travel footprint is the lodging that they choose. Hotels, motels, bed and breakfast, Airbnb's, or lodges all have some kind of environmental impact. However, these impacts vary widely and as green consumers, you have options as to where you stay and the impacts you may have. To help consumers, a number of organizations provide benchmarking and third-party verification for lodging facilities.

Most lodging verification schemes involve measuring a range of indicators associated with how the facility is managed. For example, the State of Florida has a green lodging verification program that includes indicators for indoor air quality, water conservation, energy efficiency, waste reduction and recycling, and education and communication with customers, employees, and the public (Florida DEP, 2020). Using these indicators, the Florida Green Lodging Program gives properties a set number of points and the points are used to rate properties in different categories. Across the state, there are almost 400 individual properties that have been certified in this program, including some of the largest and most well-known properties in the state including Disney's Animal Kingdom Lodge, the Hilton Orlando Lake Buena Vista in the Walt Disney World Resort, and the Four Seasons Hotel in Miami.

The US EPA provides a list of a number of other green lodging rating systems including ones by the World Travel and Tourism Council, the Global Sustainable Tourism Council, the Green Hotels Association, Audubon International Green Lodging Council, and several others (EPA, 2020). Each of these organizations provides some form of guidance as to best practices in a variety of areas including cleaning supplies, room decorations, laundry, energy use, pesticides, grounds care, food services, communication, waste management, and interaction with stakeholders.

Of course, as users of lodging, we can make a range of decisions that have big impacts on the footprint of our travel. How we eat, bathe, use sheets and towels, and a range of other micro-decisions can have big impacts on communities. For example, if you live a green life at home, but change behaviors on vacation, you can produce more waste and pollution and require more energy and water than you normally would. This impacts the communities in which you stay in deleterious ways and leads to environmental equity issues where you can impose the impacts of your bad environmental decisions on others.

Green Golfing

Golfing is not a sport that is normally seen as a particularly green one. It requires a great deal of heavily landscaped land—land that often needs water, pesticides, and herbicides to maintain grasses in very manicured ways. Golf courses are also built in inappropriate locations where they do not fit within the natural ecology of a region. There are golf courses in desert areas like Las Vegas and Saudi Arabia and in tropical areas like Vietnam and Brazil. These places have to bend their ecosystems to provide space for a sport that evolved in the cool and wet landscapes of Scotland.

In recent years, there has been a push to green the golfing industry. This initiative evolved out of growing evidence of the deleterious impacts of heavy pesticide, fertilizer, and herbicide use in the golfing industry on waterways, groundwater, and broader ecosystems. Plus, the public is demanding greener approaches to business overall and the golfing industry is responding to consumer demands. In addition, golf courses, as park-like environments, often provide home and refuge to a variety of plants and animals if they are managed appropriately.

Many golfing organizations around the world are seeking to find ways to improve their operations to embrace sustainability. For example, the International Golf Federation (IGF) states that they have made sustainability a central value in their mission listed below (IGF, 2020).

The IGF and its member organizations will:

- Make sustainability a central pillar of our mission
- Expand awareness and encourage action among golfers and golf facilities
- Promote best practice to minimize our sport's consumption of non-renewable natural resources
- Assist golf facilities to incorporate sustainable principles, practices, and technology into daily business decisions and operations
- Encourage golfers to embrace environmentally sound practice in course preparation
- Conduct high-profile golf events in an environmentally responsible manner
- Continue to work to raise the profile of our sport's progress and contribution to environmental issues
- Embrace measurement, target setting, transparency, and verification

Success in our joint efforts to promote a sustainable future for golf will deliver:

- For the game: improved financial performance and enhanced profile
- For the golfer: quality playing surfaces, value, challenges, and rewarding experiences
- For the environment: conservation of resources and biodiversity enhancement
- For the community: employment, recreational green space, and educational opportunities
- The future will present many challenges but the IGF and its member organizations are working to ensure that many more generations will enjoy golf and the facilities on which it is played.

The statement, which is rather bold for a relatively traditional field, provides a very strong foundation for understanding how the organization can help be a transformative agent for individual courses. The move to sustainability is assisted by the Golf Environment (GEO) Foundation, which seeks to drive golf to be a benefit for nature and for communities.

This group provides benchmarking tools such as third-party certification for sustainability of golf courses and a management platform that prioritizes sustainability. In addition, the group manages a climate fund to support climate-neutral golfing and golf events. As consumers, we all have choices in how we spend our recreational dollars and if you are someone who enjoys the sport of golf, you can easily find a GEO Foundation-certified course through their directory (Geo Foundation, 2020).

Adventure Tourism

One new area of tourism that has taken off in recent years is adventure tourism (Fig. 14.2). This is a form of tourism that focuses on personal adventure. Geared mainly to the young, adventure tourism provides participants with the opportunity to take some type of risk during their

Fig. 14.2 Adventure tourism is an exciting form of travel that can take you in new directions. But how green is it?

vacation time. This risk may involve mountain climbing, scuba diving, remote hikes, or visits to areas of conflict. Those who enjoy adventure tourism are in it for the thrill of the experiences. Adventure tourists often take extreme risks such as climbing a mountain wall freehand or skiing down a risky unmarked trail. They may also travel to places where they put themselves at great personal risk due to social conflicts such as travel to Iran or Afghanistan.

It must be pointed out that adventure tourism is highly focused on the personal experience and not the environment. Thus, adventure tourists who climb the mountain freehand are not doing it to experience nature; they are doing it for a thrill. Indeed, many adventure tourists do tremendous damage to the environment in their quest for the next great thrill. The growth in adventure mountain biking, for example, has led to the development of miles of pathways in mountains and deserts where they have done tremendous damage to local ecosystems.

Adventure tourism also has other costs to the public. A great example of this is the trek that many made to the bus made famous in the poignant book, *Into the Wild* (Krakauer, 1997). For those of you unfamiliar with the book, it reviews the life story of Christopher McCandless' efforts to escape modern society. He was tired of the consumerist and corporatized world he encountered through his parents and his university experiences and dropped out to become a bit of a drifter committed to seeing the world and living more authentically in harmony with nature. He eventually made it to Alaska where he hiked out to the wilderness in Denali National Park and took shelter in a bus. Sadly, he passed away in the bus—probably from eating a poisonous plant he thought was safe to eat. After his death, and the publication of *Into the Wild*, the bus became a bit of a pilgrimage for adventure tourists interested in following a similar path as Chris McCandless.

Unfortunately, the trek to the famous bus is dangerous and rescue teams have rescued many who got lost on the trail or who found themselves in dangerous conditions. Plus, two adventurers drowned while trying to make the trek. As a result of these problems, the State of Alaska removed the bus to avoid the costs and trouble associated with rescuing people trying to reach it. It will eventually be on display at the Museum of the North at the University of Alaska at Fairbanks.

What the Alaska example points out is that adventure tourism is really about the self. The growth of Instagram and YouTube platforms helps to create audiences for adventure tourists who take great risks to satisfy their followers. The number of people taking extreme risks is growing—often with tragic outcomes. For example, three young adventurers all died when they fell 30 meters off a waterfall in British Columbia (Wootson Jr., 2018). They were YouTube stars that took great risks and posted them to their High on Life YouTube channel which had 500,000 subscribers and their Instagram account which had over a million subscribers.

Ecotourism

While some tourists like to take extreme risks regardless of environmental consequences, there are other ecotourists who like to travel to places to experience nature without doing any harm to it. A whole field of ecotourism has blossomed in the last few decades as travelers have sought to have authentic experiences responsibly. Ecotourism has blossomed especially in tropical zones where tourists are able to come into intimate contact with biodiverse regions. Ecotourism is also an important activity in the grasslands of Africa where ecotourists come to see large mammals like the African elephant and rhino. Some areas of the extreme north also draw tourists to see the northern lights. Many local areas try to promote ecotourism by promoting hiking, fishing, or other activities.

One important aspect of ecotourism that is different from other tourist activities that may involve experiences with nature is that there is an intention within the field to promote some type of economic development for local communities. Thus, ecotourism is not just about seeing nature, it is also about helping local areas pull out of poverty through some action. In addition, ecotourism seeks to promote preservation of landscapes or ecosystems. For example, the safari tourism in Africa helps to preserve land and keep the large animals safe.

There is one subset of ecotourism that is a bit troubling—doom tourism. This is tourism that seeks to see places before they are destroyed by climate change or other human activities. Doom tourists have been going to see the alpine glaciers that are disappearing and have been visiting reefs as they are in decline.

It must be pointed out that ecotourism is sometimes seen by some as virtue signaling and not an authentic form of sustainable tourism. People have to travel great miles by air to get to the tourist site where they will certainly have some type of impact on people and the environment. While there certainly is a streak of naivety in ecotourism about its impacts, there is no doubt that this form of tourism usually has a better outcome than traditional tourism.

Volunteer Tourism

Volunteer tourism is somewhat related to ecotourism in that people engage in it to try to feel like they are making a difference. There are a range of volunteer travel types. At the most basic level, church missions that leave the developed world to go help a church in a developing country is an example. However, there are highly organized trips that are put together by non-profit and for-profit organizations that seek to create volunteer experiences around the world. Some of these experiences involve improving ecosystems such as reef restoration, tree planting, or some form of environmental clean-up.

The critiques of volunteer tourism note that this is a highly intrusive form of tourism that can do damage to local communities. They note that it can be easily seen as a type of colonialism in that the volunteers are largely from Western countries that are trying to transform places based on Western values and norms. Plus, local communities, that are often struggling, have to make space for outsiders and often provide high levels of resources to support their desires during their volunteer experiences. In addition, these "parachute in, parachute out" types of experiences may not actually leave long-standing change in a local community.

How You Can Transform Your Vacations and Travel

More and more of us want to make a difference through our travel and tourism dollars and this chapter outlined several ways to accomplish this goal. Table 14.1 provides a summary of how you can green your travel and green your vacation. At the most basic level you can green your travel by purchasing carbon credits for travel. This helps to make each trip carbon neutral so you do not advance the trend of doom travel. A simple way to green a vacation is to make sure that you use only certified green travel vendors. These include green hotels, green cruise lines, and green-certified golf courses.

You can make some harder or more expensive choices in travel by committing to taking trains or buses when you travel. These trips can take more time, but they are far better for the environment. You could also commit to using electric cars whenever possible. On vacation, you can explore ecotourism options that are authentically green. To deepen your commitment to sustainability, you can commit to cutting all travel by airplane and by getting rid of your car. You can rent a car—electric of course—only when necessary. To fully green your vacation time, you can

Table 14.1 Simple, hard/expensive, and innovative/life-changing approaches to greening your travel and vacations

	Ways to green your travel	Ways to green your vacation
Simple	Purchase carbon credits for your travel—especially for car and air travel	Make sure that you travel and use only certified green travel vendors such as hotels, cruise lines, and golf courses
Hard/expensive	Commit to taking some form of mass transit (bus or train) for your travel Use an electric vehicle	Explore ecotourism options that are authentically green
Innovative/life-changing	Commit to cut all travel by airplane Only rent a car (electric) when absolutely necessary	Stay local and vacation in your region

stay local and explore your region. By doing so you will deepen your understanding of your own world and also significantly cut your carbon footprint.

References

EPA. (2020). Green hotels—Resources, ecolabels, and resources. Retrieved December 5, 2020, from https://www.epa.gov/p2/green-hotels-resources-ecolabels-and-standards

Florida DEP. (2020). Green Lodging. Retrieved December 2020, from https://floridadep.gov/OSI/Green-Lodging

Geo Foundation. (2020). Directory. Retrieved December 5, 2020, from https://sustainable.golf/directory

IGF. (2020). IGF sustainability statement. Retrieved December 5, 2020, from https://www.igfgolf.org/sustainability/igf-statement-on-sustainability

Krakauer, J. (1997). *Into the Wild*. Villard.

Solar Impulse Foundation. (2020). Solar Impulse Foundation. Retrieved December 5, 2020, from https://solarimpulse.com

Wootson Jr., C. R. (2018, July 7). YouTube daredevils filmed dangerous stunts for clicks—Then died going over a waterfall. *The Washington Post*. Retrieved December 5, 2020, from https://www.washingtonpost.com/news/the-intersect/wp/2018/07/07/youtube-daredevils-filmed-dangerous-stunts-for-clicks-then-died-going-over-a-waterfall/

15

Tune Out, Buy Nothing, and Get Educated

Introduction

Our current era is full of distractions that have limited real use if you choose to live an authentic life. An authentic life is one in which we have strong personal relationships with family and friends while living according to our personal values. Since you are reading this book, environmentalism and sustainability must be part of your own ethic. Indeed, given the range of problems our planet is facing, many of us have developed some sort of green ethic within our consciousness and struggle with finding ways to make it a greater part of our authentic self (Fig. 15.1).

Most of us would like to live authentic lives but get sidetracked by a range of distractions. A variety of things can take us off the path of living authentically. For example, we can overuse social media or television. By doing so, we are influenced by others in ways that may not be the best for us. There are no filters on social media and it can be a morass of negativity and lies. Professional influencers can push you into consumerism or a belief system that is incongruent with your authentic self. This final chapter takes a look at some of the issues we are facing in society due to an overwhelming culture of consumerism and misinformation. It also

Fig. 15.1 A group of community volunteers installing an osprey nest structure on the shores of Long Island Sound

reviews how we can overcome these influences via a concentrated effort of immersion in environmental and sustainability education while making sound lifestyle choices in order to live a more authentic life with our family and friends.

Consumerism and Social Media

When *Keeping Up with the Kardashians* premiered in 2007, social media was relatively new. Twitter started in 2006 and Facebook premiered in 2004. Instagram was not a thing until 2010. The Kardashians series was a big hit with young people and the Kardashian family became major social media stars as they posted pictures of themselves with celebrities and with the various consumer goods they acquired. They ushered in a whole new world of social media that focused attention on how to influence others. Soon, the idea caught on and many celebrities became very active on social media and became influencers in their own right.

Today, there are social media influencers from a number of different fields. Certainly, celebrities like Kylie Jenner and Selena Gomez have huge numbers of followers, but so do political leaders like Joe Biden and Donald Trump. Even scientists like Michael Mann and Bill Nye have huge followings. When people like this post Tweets or Instagram posts, people notice. This level of access to famous people is relatively new. It has changed our culture tremendously and changed how companies interact with consumers. They can pay a celebrity social media star to post photos with their products and reach a bigger audience than they could ever reach with magazine or billboard advertisements in today's world.

Companies that promote consumerism have gotten very good at using social media. However, the scientific and environmental community is a bit late to the game. While the number of scientists on social media platforms is growing, they do not have the reach of others who are driving overconsumption and bad personal environmental decision-making. Social media is more and more used for a variety of marketing purposes that focuses not only on consumerism but also on political ideology. A new profession of social media influencers has emerged since social media erupted in our culture. Some of the influencers are individuals, like Kylie Jenner, who promote their brand or other products while some are nameless individuals working for companies that flood social media with a particular product or idea. For example, social media bots account for about 10% of Twitter content (Imperva, 2020). These bots automatically generate social media content to influence consumers or voters. The 2016 US election may have been influenced by bots. Because of the significant amount of lies in the bot-driven content, Twitter and Facebook have been under tremendous pressure to remove or tag falsehoods.

One great example of the power and problem of social media is to look at the climate change issue. There really are only a handful of scientists who do not recognize that humans have significantly altered the climate and they are largely marginalized within the scientific community. Many of the papers or posters at the last Geological Society of American meeting I attended dealt in some way with human-induced planetary change including climate change. However, if you spend any time on Twitter or Facebook and look at climate change discussions, you would think there was a raging debate in the scientific community. There is not. That train has left the station a decade ago. What remains is a group of people fighting a losing battle who are highly influenced by those who have the most to gain from continuing to use fossil fuels.

Similar social media "debates" occur around a range of other sustainability issues such as deforestation, environmental justice, recycling, and pollution. Almost any issue that is a no brainer in terms of the science or policy issue becomes twisted or changed online. I remember going to a public meeting about the development of a green streets program in my community several years ago. During the public comment section, a speaker came to the podium and denounced the program as part of Agenda 21 which was taking over local governments. Agenda 21, of course, came out of the Earth Summit in Rio in 1992 and urged local governments to take action on sustainability in some way. ICLEI for Local Governments was the vehicle for helping local governments develop goals around sustainability. However, there was no mandate and there certainly is not widespread acceptance of local sustainability programs in the United States or most other countries. The green streets program in my community emerged from local leaders seeking to improve traffic flow and walkability in the community's downtown. The green streets program passed, but after the meeting I did ask the speaker why she thought the program was part of Agenda 21 and an effort by the United Nations to control local governments. She told me she read about it on Facebook.

The problem with social media has gotten so bad that many are turning it off. Millions of users have left Facebook due to what many see as bad information encouraged by bad Facebook policies. New platforms, like TikTok, have emerged, but there is growing concern about how these companies mine data on their millions of users. The information is used to influence us as consumers and as voters. We are entering into a new world where social media can tell how much time we are spending in different places and with whom we are spending time. This intrusion into our lives is accomplished using our smartphones which are near constant companions for many of us.

To counteract this intrusion, many are not only rejecting social media, but also smartphone technology. They are switching to old-fashioned flip phones that do not have the GPS technology and extensive apps options. They are using a phone purely for its original intended use: communication with loved ones and friends. While these people may have used social media at some time, they are finding that giving up the distraction and technology helps them live more authentic lives.

Consumerism, driven in part by social media, influences many decisions as to how we live our lives. It is worth considering three areas of consumerism—fast fashion, electronics, and food—to examine the impact on global systems. Fast fashion, which has been discussed elsewhere in this book, involves the production of cheap clothing, jewelry, and accessories within the global supply chain. Clothing seasons of the past do not matter. Now, there is a constant barrage of updated looks throughout the year. Many people consider clothing that is worn only once or twice to be old (Siegle, 2019). Each American throws away nearly 100 pounds of clothing each year (Gilmore, 2018). The quest for new looks is driven by irrational thinking that one needs to have constant new looks to be acceptable in society. This drive for freshness is of course driven by social media and is leading to significant environmental and social ills.

Our electronic use is another form of consumption driven by social media. We want to have the latest, coolest smartphone with all of the bells and whistles or the latest gaming system. We have made many mundane things "smart" and connected to the Internet. Our security systems, our watches, our home heating systems are all connected. We have an Internet of things that is providing constant updates on when our groceries will be delivered and how our kids are doing on their homework. Thus, the issue with electronics is not just the hardware, but the vast amount of information that is readily available to us. For many people, this level of access is overwhelming and concerning. Of course, our access to electronics also provides access to a range of social media platforms which also drives us to purchase new electronics. Of course, as pointed out elsewhere, this quest for more and more electronics has significant environmental and social costs.

Our desire for unique food is another driver of consumerism. We want all kinds of food available to us, whenever we want it, regardless of season. Want a fresh peach pie in Scotland in January? No problem! Our global food network can provide them for you in your grocery store. We have created a global food system that has lost touch with the local and which fetishizes the exotic. We all love to look at social media to see what people are cooking or eating. In addition, food and travel channels and shows like the Great British Bake Off or Chopped urge people to try new recipes and purchase new kitchen implements. While it may seem like there is nothing significantly wrong with any of this, it is important to point out that our current system leads to tremendous amount of food waste and that our current food production and distribution system is unsustainable.

How to Change?

The big question, of course, is how do we change? How do we all get off of the unsustainable treadmills of our society? The answer lies in individual change and leadership. There are three main ways we can foster significant change in our lives and in the lives of others. These involve lifestyle changes, leadership, and education. Each of these issues will be discussed below.

Lifestyle Changes. Throughout this book, I have tried to provide guidance as to how you can make big and small decisions to transform your life into one that is more sustainable. All of these decisions involve some type of lifestyle changes. Some of these are small, such as embracing meatless Mondays, while others are larger, such as committing to living without a car. However, the key point is that all of these changes are possible if you truly embrace an environmental and sustainability ethic.

The thing about ethical living around sustainability is that it involves making big planned decisions and small micro-decisions. In many ways, the big, planned decisions are the easy ones. You can plan out how to take a green vacation or how to develop a solar energy system for your home.

However, it is more difficult when you are confronted with day-to-day decision-making where you have to make a quick decision around something that could have a big environmental impact. That is why it is important to be very clear in your thinking around your sustainability ethic so that it is easy for you to make wise snap judgments. For a simple example, you may be at a restaurant confronted with a range of menu choices that include things like steak, hamburgers, fish, and vegan pasta primavera. You have a range of choices that you could make based on your desire for something delicious. But, from an ethical frame of mind, you know that the vegan dish is better for the planet.

These small decisions add up and will help you live a much greener life. When seeing a sale for clothing online, you know that even if the sale might be great, you have plenty of clothing. When thinking about throwing out a broken washing machine, you know it could be fixed or sold for parts. When choosing how to manage your lawn or garden, you know you have options other than chemical pesticides and fertilizers. These micro-decisions matter and help keep your ecological footprint low.

The lifestyle changes are not just about how you spend your money but also about how you spend your time. If you spend your time consuming media that promotes unsustainability you will likely live a less sustainable life. For example, if you spend any time on YouTube or other social media sites, you will be confronted with a range of advertisements that try to sell you things based on your browsing history. Television shows and movies use product placements to make you think about purchasing a product. In our modern world we are bombarded with advertisements and influencers who try to make us purchase things or think a certain way. We give up a bit of control when we consume these media sources. Spending less time in front of a computer, television, or smartphone helps us focus on our own life and our place within the planet.

Most of us with smartphones spend over two hours a day looking at the screen. Think of all of that wasted time! It is so easy to get sucked into the Internet on our smartphones, particularly when there is interesting breaking news. With all of the odd and bad news in 2020, particularly related to the global pandemic, doomscrolling became a new word to refer to the act of searching the Internet, particularly Twitter, for updates on bad situations. Oxford English Dictionary, which normally picks one word of the year, picked several, including doomscrolling, as words of an unprecedented year for 2020 (Oxford English Dictionary, 2020). While many people spent the 2020 pandemic doomscrolling, others turned off their phones and learned new skills, created amazing pieces of art, and read great literature. We all have choices as to how we spend our time and just as we can make micro-decisions around how we eat or make purchases, we can also choose to spend time more authentically with our family and in the betterment of ourselves and others.

Sustainability is also about committing to transforming your community into a more sustainable place. Many of us spend so much time online that we do not know what is happening in our community or neighborhood. By turning off, we can meet new friends, find new causes to care about, volunteer to help others, go to public meetings, and find ways to improve the environment around us. By spending so much time online, we have created virtual communities that have very little to do with the people and environments around us that actually impact our real daily lives. We can choose to become more focused on our real community that is always around us than our virtual communities that are fleeting in nature.

Leadership. Many of us are followers and few of us are leaders. It is hard to take responsibility for an initiative or project. Throughout our early lives we are taught to be followers. Our parents tell us what to do and when we got to school, we got tons of direction on everything from the amount and type of homework we get to the time we are allowed to use the restroom. In adulthood, we follow direction from our employers and follow well-trod paths for how to live. It is relatively easy to live a quiet life without much fanfare or initiative outside of our work and family obligations. However, some of us reach out of our comfort zones to get engaged with issues that matter to us in our community, our region, our state, and even the world. In doing so, we provide leadership on topics that matter to us. From the context of this book, the topic that is pertinent is sustainability and environmentalism.

To become an effective sustainability leader, it is important to have a vision as to your values and what you want to accomplish. What is the core ethic that is driving you to step up and lead? Is it broad environmentalism and protection of the planet? Is it a drive to stop litter? Is it environmental justice? By understanding the ethical underpinnings of your initiatives, you can better articulate your vision to others. Indeed, it is very helpful to develop a vision statement such as "My vision is to increase environmental awareness in my community and to preserve natural habitat."

By creating a clear vision statement, you help to define your focus. One of the most important aspects of a vision statement is that you find out what you are not doing. Some leaders try to do too much and their work is then diluted. The vision statement above is very clear and does not do things like work on transportation issues or green energy. Instead, the leader is all about environmental awareness and land preservation. Clarity in leadership matters and the clearer you are to your stakeholders, the more support you will have on targeting goals. In fact, a vision statement can help articulate goals for you and for any group of people you may lead. Using the example above, two sets of goals can be articulated. One set can focus on environmental awareness and one set can focus on land preservation. The awareness goals may include things like conducting public education events, developing a website or blog, creating social media content, or creating signage. The preservation goals may include things like ecosystem restoration of public lands, fundraising and purchasing of sensitive lands, or protection of threatened or endangered species. The point is that you can develop a clear set of goals around your vision.

You can build support for your goals by engaging stakeholders. As a leader working on a particular issue, it is important to build support for your goals with others. If you are working locally, you can network with local political or business leaders and with others who are advocating for environmental or sustainability issues. You can attend local meetings and find ways to discuss your issues. You could partner with groups that have similar visions to achieve your goals and even join their leadership team. You could even start a non-profit group to give yourself a broader platform.

If you are working on broader regional, national, or international goals, such as protecting coral reefs around the world, there are many groups working on similar issues and it will be important to network with the groups and find ways to help through your own leadership initiatives. For example, if your broad vision is to protect coral reefs, you could help fundraise to restore reefs or advocate with political leaders to preserve reef habitat.

No matter what your goals might be, it is important to put in the work to try to achieve them. There are lots of "feels" about the environment and sustainability. People want to protect the planet and make the planet more sustainable, but they do not often want to do anything about these issues. They may Tweet about it or do a Facebook post, but this form of activism, aptly named slacktivism, does not have much of an impact. Leaders put in time on their projects. In order to get results, you have to be willing to work for your goals.

There are many environmental leaders from whom we can draw inspiration. Margaret Kenyatta, for example, is the first lady of Kenya who is known for encouraging simplicity of lifestyle and environmental protection (van Wyk, 2018). She has advocated for a number of environmental causes including strong support for environmental education. As the first lady of one of Africa's most important countries, she could have opted for an opulent lifestyle. Instead, she lives simply, promotes sustainability, and advocates for education.

Another leader who has already been mentioned in this book is Greta Thunberg. As a young girl, she recognized that political leaders were not doing enough to stop the myriad of problems associated with global climate change (Thunberg, 2018). She joined forces with other youth on the Fridays for Future initiative which included well-publicized climate strikes. She has inspired a generation of youth through her strong voice and personal commitment.

Margaret Kenyatta and Greta Thunberg are examples of global leaders. However, there are thousands and thousands of local leaders who are doing amazing things in their communities. For example, Adrienne Esposito, who runs the Citizens Campaign for the Environment on Long Island, is deeply focused on improving surface and groundwater across the highly developed island where residents depend on a single aquifer for drinking water (Citizens Campaign for the Environment, 2020). One major pollution event can wreck the aquifer for hundreds of thousands of people. Esposito has focused like a laser on educating a range of stakeholders on issues of water quality and has advocated with local, state, and national political leaders to press for improved rules and regulations to protect water across Long Island. Each region has leaders like Esposito who focus on local issues. They identify a problem, develop goals as to how to address the issue, and then put in the hard work to reach their goals.

It is important to note that many people are hesitant to get into environmental leadership roles because they feel that they will not be taken seriously because their lifestyle is not "green" enough. This is a longstanding problem in the environmental community. There is a sense that to be truly "green" you have to live the total green lifestyle with solar panels, electric cars, small houses, and so on. The reality is that very few environmental leaders live that lifestyle. They live like regular people in their communities. Certainly, they likely have adopted aspects of a green lifestyle, but they do not necessarily have all of the characteristics of the stereotypical environmentalist—whatever that is. What matters the most, however, is commitment, passion, and understanding. Greta Thunberg does not have a college degree (yet) but she does have a drive to make a difference in the world. She is an environmental leader because she stood up and started taking action to make the world a better place.

Education. While you do not need an education to change your lifestyle or become a leader, it helps. There are many ways to get educated on sustainability and environmental issues (Fig. 15.2). Perhaps the best way to become familiar with environmental issues is to read a range of materials. The fact that you are reading this last chapter in this book shows that you are already taking on this challenge. However, there are many great sources of sustainability information. At the local level, it is a great idea to read your local newspapers. They often provide a range of information that matters to your community. In my area, there is often information about local water issues or development plans that contributes to my knowledge. In addition, the local newspaper has editorial content where you can get a sense of local opinion on environmental issues and where you can contribute your own opinions as well. At the national or world level, great newspapers like *The New York Times* provide a range of timely content.

You can also find great sustainability content on blogs and podcasts. My blog, *On the Brink*, captures news content and often features series about my own publications, national parks and monuments, or

Fig. 15.2 Celebrating Earth Day, like these students, helps to build community and participants can share new sustainability strategies

important sustainability topics like the Dakota Access Pipeline. The podcast, *The Minimalists*, features great conversations on how to live a more minimalist life. No matter your interest in sustainability, you can find blogs and podcasts that will allow you to take deep dives into content that matters to you.

Government websites also have a great deal of sustainability information that ranges from technical reports to data. For example, your local government probably has published some type of sustainability plan which lists a range of priorities for the region with distinct sustainability goals. Technical reports on local issues like water resources, water or air quality, environmental justice, and economic development are great sources of knowledge about your region. You can also find records of meetings where environmental issues were discussed or decided. These sites also contain the regulations for the entity being governed such as building codes, air quality standards, and food safety standards. National government sites also contain similar types of information. However, they often provide guidance for best practices. For example, the US

Environmental Protection Agency provides information for how to engage stakeholders to create improved environmental outcomes.

Websites of non-profit organizations and businesses also contain a great deal of information. For example, if you are interested in green building, the website of the US Green Building Council (USGBC), a non-profit organization that promotes sustainable buildings, contains a huge amount of information about their rating system and how to make buildings more sustainable. There are local chapters of the USGBC and their websites provide information pertinent to your community.

As mentioned in Chap. 13, most major businesses are deeply engaged with sustainability issues. Their websites provide ways for them to communicate their initiatives with their stakeholders. For example, the giant retailer, Target, has a whole section on their website under the heading of Corporate Responsibility. Within the Planet section, there is information about climate, chemicals, deforestation, water, sustainable operations, sustainable products, and responsible sourcing. For example, under the climate heading, Target lists the goal of reducing greenhouse gas emissions by 30% by 2030 based on emissions as of 2017. In addition, they are requiring suppliers to cut Scope 1 and 2 emissions by 80% by 2023. They outline ways that they will achieve these goals by investing heavily in green energy and by improving lighting efficiency. The amount of sustainability information available on corporate websites is really rather astounding and is far different from just a decade ago when many corporations were not particularly engaged on sustainability and climate change. There is a new culture of corporate responsibility that involves providing information to the public about their sustainability outcomes.

Of course, books are one of the best sources of sustainability information. There are a number of organizations that have listed the best sustainability books ever written. For example, Cambridge University has listed their top 50 sustainability books ever published that include Rachel Carson's *Silent Spring* and Ralph Nader's *Unsafe at Any Speed* (Cambridge Institute for Sustainability Leadership, 2020). While many of these books are definite classics, some may be a bit dated in content. For more contemporary books, you can look for new releases or best sellers. For example, the book *Braiding Sweetgrass: Indigenous Wisdom, Scientific Knowledge, and the Teachings of Plants*, by Robin Wall Kimmerer, is a book about plants and ecology as seen from a Native American perspective that is currently on the *New York Times* list of bestselling books in paperback nonfiction (Krimmerer, 2013). You can also explore titles from particular publishers such as the one that is publishing this book.

Some of my favorite places to look for sustainability and environmental knowledge are the websites of the university presses. Most large universities around the world operate publishing houses that produce books produced by the academic community that would be of interest to a larger audience. Some of the presses focus on regional content or specialize in some particular area. One press that has published some of my writing, the University Press of Florida, publishes dozens of books each year. Their most recent catalog includes books on native plants of Florida, suburban sustainability, and archaeology and ecology in Dominica. It is worth taking a look to see what your local university press is publishing to get a better understanding of the kind of work being published in your region by the leading voices in your community.

There are also ways that you can gain formal education in sustainability and environment. Most colleges and universities offer courses or degrees on environmental and sustainability topics. Sustainability is a growing field and there is more demand for experts with particular training in the area. There are many universities that offer distinct degrees in environmental science, environmental policy, environmental studies, or environmental law. The options are almost limitless and you can probably find a degree somewhere that matches your interests. Many K-12 schools also offer some type of environmental curriculum and some schools have made sustainability a key component of their educational mission.

Putting It All Together

This book has provided a range of information to help you live a greener life, change your community, and transform your organizations. Each of us has the ability to help to change the direction of our society so that we stop doing damage to the planet. At this moment in our history, we are doing more ecological transformation than any other time on our planet.

We are seeing mass extinctions, changing climates, and degraded waterways. We are seeing greater social ills, huge income disparities, and racism. Yet, we can each make decisions to make things better.

We can each change our lifestyles to help make our planet more sustainable and equitable. We can also step up and be leaders to help move the positive change forward. We can also improve our sustainability knowledge by consuming a variety of sources of information. The future is in our hands. We either can continue to move forward on our path of planetary destruction or can change direction now. To me, the answer is simple. We each need to be the change we want to see. In so doing, we will help to transform the world into one that values nature, protects the environment, and supports a just and equitable society.

References

Cambridge Institute for Sustainability Leadership. (2020). *The top 50 sustainability books*. Retrieved December 5, 2020, from https://www.cisl.cam.ac.uk/resources/archive-publications/the-top-50-sustainability-books

Citizens Campaign for the Environment. (2020). Citizens Campaign for the Environment. Retrieved December 5, 2020, from https://www.citizenscampaign.org/

Gilmore, N. (2018, January 16). Ready-to-waste: America's clothing crisis. *Saturday Evening Post*. Retrieved December 5, 2020, from https://www.saturdayeveningpost.com/2018/01/ready-waste-americas-clothing-crisis/

Imperva. (2020). Bots. Retrieved December 5, 2020, from https://www.imperva.com/learn/application-security/what-are-bots/#:~:text=Social%20Media%20Bots,Twitter%20accounts%20are%20social%20bots

Krimmerer, R. W. (2013). *Braiding sweetgrass: Indigenous knowledge, and the teachings of plants*. Milkweed Editions.

Oxford English Dictionary. (2020). Word of the year. Retrieved December 5, 2020, from https://languages.oup.com/word-of-the-year/2020/

Siegle, L. (2019, June 21). Fast fashion is on the rampage, with the UK at the head of the charge. *The Guardian*. Retrieved December 5, 2019, from https://www.theguardian.com/fashion/2019/jun/21/fast-fashion-is-on-the-rampage-with-uk-at-the-head-of-the-charge

Thunberg, G. (2018). *No one is too small to make a difference*. Penguin Books.

van Wyk, J. (2018). Political leadership and sustainability in Africa: Margaret Kenyatta. In R. Brinkmann & S. J. Garren (Eds.), *The Palgrave handbook of sustainability: Case studies and practical solutions* (pp. 347–358). Palgrave Macmillan.

References

350.org. (2020). *350.org*. Retrieved December 5, 2020, from https://350.org
AASHE. (2020a). The association for the advancement of sustainability in higher education. Retrieved December 5, 2020, from https://www.aashe.org
AASHE. (2020b). The sustainability tracking, assessment & rating system. Retrieved December 5, 2020, from https://stars.aashe.org
Abbey, E. (1975). *The monkey wrench gang*. Lippincott Williams & Wilkins.
Albritton, R. (2019). *Eco-socialism for now and the future: Practical utopias and rational action* (Palgrave Insights into Apocalypse Economics). Palgrave Macmillan.
Altieri, A. H., & Gedan, K. B. (2015). Climate change and dead zones. *Global Change Biology, 21*, 1395–1406.
Anderson, R. C., & White, R. (2009). *Confessions of a radical industrialist: Profits, people, purpose—Doing business by respecting the earth*. St. Martin's Press.
Anderson, T. R., Hawkins, E., & Jones, P. D. (2016). CO_2, the greenhouse effect and global warming: From the pioneering work of Arrhenious and Callendar to today's Earth System Models. *Endeavor, 40*, 178–187.
Apple. (2020). Apple commits to be 100 percent carbon neutral for its supply chain and products by 2030. Retrieved December 5, 2020, from https://www.apple.com/newsroom/2020/07/apple-commits-to-be-100-percent-carbon-neutral-for-its-supply-chain-and-products-by-2030/
Associated Press. (2019, May 28). Teen climate activist Greta Thunberg addresses leaders at world summit. *The Washington Post*. Retrieved June 1, 2019, from

https://www.washingtonpost.com/lifestyle/kidspost/teen-climate-activist-greta-thunberg-addresses-leaders-at-world-summit/2019/05/28/a34c7d04-7cb3-11e9-8ede-f4abf521ef17_story.html?utm_term=.e327a3aa439b

Biello, D. (2013, December 12). How nuclear power can stop global warming. *Scientific American.* https://www.scientificamerican.com/article/how-nuclear-power-can-stop-global-warming/

BlackRock. (2020). A fundamental reshaping of finance. Retrieved December 5, 2020, from https://www.blackrock.com/corporate/investor-relations/larry-fink-ceo-letter

Blum, M. D., & Roberts, H. H. (2009). Drowning of the Mississippi Delta due to insufficient sediment supply and global sea-level rise. *Nature Geoscience, 2*, 488–491.

Boer, M. M., Resco de Dios, V., & Bradstock, R. A. (2020). Unprecedented burn area of Australian forest fires. *Nature Climate Change, 10*, 171–172.

Borrelli, P., Robinson, D. A., Fleischer, L. R., Lugato, E., Ballabio, C., Alewell, C., Meusburger, K., Modugno, S., Schütt, B., Ferro, V., Bagarello, V., Van Oost, K., Montanarella, L., & Panagos, P. (2017). An assessment of the global impact of 21st century land use on soil erosion. *Nature Communications, 8*, 2013.

Brei, M., Perez-Barahona, A., & Strobl, E. (2016). Environmental pollution and biodiversity: Light pollution and sea turtles in the Caribbean. *Journal of Environmental Economics and Management, 77*, 95–116.

Brinkmann, R. (2013). *Florida sinkholes: Science and policy.* University Press of Florida.

Brinkmann, R. (2017, February 26). Dakota Access Pipeline teaching resources. *On the Brink.* https://bobbrinkmann.blogspot.com/2017/02/dakota-access-pipeline-teaching.html

Brinkmann, R. (2018). Economic development and sustainability: A case study from Long Island, New York. In R. Brinkmann & S. J. Garren (Eds.), *The Palgrave handbook of sustainability: Case studies and practical solutions* (pp. 433–450). Palgrave Macmillan.

Brinkmann, R. (2019, November 19). Brazil deforestation continues and indigenous people fight for survival amid calls for divestment. *On the Brink.* https://bobbrinkmann.blogspot.com/2019/11/brazil-deforestation-continues.html

Brinkmann, R. (2020). *Environmental sustainability in a time of change.* Palgrave Macmillan.

Brinkmann, R., & Tobin, G. A. (2001). *Urban sediment pollution.* Springer.

Brundtland, G. H. (1987). *Report of the World Commission on Environment and Development: Our common future.* https://sustainabledevelopment.un.org/content/documents/5987our-common-future.pdf

Bullard, R. D. (2000). *Dumping in dixie: Race, class, and environmental quality* (3rd ed.). Westview Press.

Burney, D. A., & Flannery, T. F. (2005). Fifty millennia of catastrophic extinctions after human contact. *Trends in Ecology & Evolution, 20*, 395–401. https://doi.org/10.1016/j.tree.2005.04.022

Cambridge Institute for Sustainability Leadership. (2020). *The top 50 sustainability books.* Retrieved December 5, 2020, from https://www.cisl.cam.ac.uk/resources/archive-publications/the-top-50-sustainability-books

Carbon Footprint. (2020). *Carbon calculator.* Retrieved December 5, 2020, from https://www.carbonfootprint.com/calculator.aspx

Carson, R. (1962). *Silent spring.* Houghton Mifflin.

Chen, J., Qian, H., & Wu, H. (2017). Nitrogen contamination in groundwater in an agricultural region along the New Silk Road, northwest China: Distribution and factors controlling its fate. *Environmental Science and Pollution Research, 24*, 13154–13167.

Citizens Campaign for the Environment. (2020). *Citizens Campaign for the Environment.* Retrieved December 5, 2020, from https://www.citizenscampaign.org/

Connell, J. (2016). Last days in the Carteret Islands? Climate change, livelihoods, and migration on coral atolls. *Asia Pacific Viewpoint, 55*, 3–5.

Constable, A. L. (2016). Climate change and migration in the Pacific: Options for Tuvalu and the Marshall Islands. *Regional Environmental Change, 17*, 1029–1038.

Costanza, R., Mitsch, W. J., & Day, J. W. (2006). A new vision for New Orleans and the Mississippi delta: Applying ecological economics and ecological engineering. *Frontiers in Ecology and the Environment, 4*, 465–472.

Crutzen, P. J. (2006). The "Anthropocene". In E. Ehlers & T. Krafft (Eds.), *Earth system science in the Anthropocene* (pp. 13–18). Springer.

Cutter, S. L. (2006). *Hazards, vulnerability and environmental justice.* Earthscan.

Davidson, N. C. (2014). How much wetland has the world lost? Long-term and recent trends in global wetland area. *Marine and Freshwater Research, 65*, 934–941.

Douglas, M. S. (1947). *The Everglades, river of grass.* Rinehart & Company.

Dris, R., Imhof, H., Sanchez, W., Gasperi, J., Galgani, F., Tassin, B., & Laforsch, C. (2015). Beyond the ocean: Contamination of freshwater ecosystems with (micro-) plastic particles. *Environmental Chemistry, 12*, 539–550.

Ekstrom, J. A., Suatoni, L., Cooley, S. R., Pendleton, L. H., Waldbusser, G. G., Cinner, J. E., Ritter, J., Langdon, C., van Hooidonk, R., Gledhill, D., Wellman, K., Beck, M. W., Brander, L. M., Rittschof, D., Doherty, C., Edwards, P. E. T., & Portela, R. (2015). Vulnerability and adaptation of US shellfisheries to ocean acidification. *Nature Climate Change, 5*, 207–214.

Enerdata. (2020a). Coal and lignite domestic consumption. Retrieved December 5, 2020, from https://yearbook.enerdata.net/coal-lignite/coal-world-consumption-data.html

Enerdata. (2020b). Oil products domestic consumption. Retrieved December 5, 2020, from https://yearbook.enerdata.net/oil-products/world-oil-domestic-consumption-statistics.html

Enerdata. (2020c). Natural gas domestic consumption. Retrieved December 5, 2020, from https://yearbook.enerdata.net/natural-gas/gas-consumption-data.html

EPA. (2019). Sources of greenhouse gas emissions. Retrieved June 1, 2019, from https://www.epa.gov/ghgemissions/sources-greenhouse-gas-emissions

EPA. (2020a). Local greenhouse gas inventory tool. Retrieved December 5, 2020, from https://www.epa.gov/statelocalenergy/local-greenhouse-gas-inventory-tool

EPA. (2020b). U.S. greenhouse gas reporting archive. Retrieved December 5, 2020, from https://www.epa.gov/ghgemissions/us-greenhouse-gas-inventory-report-archive

EPA. (2020c). How we use water. Retrieved December 5, 2020, from https://www.epa.gov/watersense/how-we-use-water

EPA. (2020d). EJSCREEN: Environmental justice screening and mapping tool. Retrieved December 5, 2020, from https://www.epa.gov/ejscreen

EPA. (2020e). Green hotels—Resources, ecolabels, and resources. Retrieved December 5, 2020, from https://www.epa.gov/p2/green-hotels-resources-ecolabels-and-standards

FIPR Institute. (2020). Mining and beneficiation. Retrieved December 5, 2020, from https://fipr.floridapoly.edu/research/mining-and-beneficiation-mineral-processing.php

Florida DEP. (2020). Green Lodging. Retrieved December 2020, from https://floridadep.gov/OSI/Green-Lodging

Florida Green Building Coalition. (2020). *Florida Green Building Coalition*. Retrieved December 5, 2020, from https://floridagreenbuilding.org

Franta, B. (2018, September 19). Shell and Exxon's secret 1980s climate change warnings. *The Guardian*. Retrieved June 1, 2019, from https://www.theguardian.com/environment/climate-consensus-97-per-cent/2018/sep/19/shell-and-exxons-secret-1980s-climate-change-warnings

Fujitsu. (2020). Green IT: The global benchmark. Retrieved December 5, 2020, from http://www.ictliteracy.info/rf.pdf/green_IT_global_benchmark.pdf

Garner, R. (1996). *Animal rights: The changing debate*. Palgrave Macmillan.

Geo Foundation. (2020). Directory. Retrieved December 5, 2020, from https://sustainable.golf/directory

Gilmore, N. (2018, January 16). Ready-to-waste: America's clothing crisis. *Saturday Evening Post*. Retrieved December 5, 2020, from https://www.saturdayeveningpost.com/2018/01/ready-waste-americas-clothing-crisis/

Gladstone, R. (2019, March 27). Cholera, lurking symptom of Yemen's War, appears to make roaring comeback. *The New York Times*. Retrieved June 1, 2019, from https://www.nytimes.com/2019/03/27/world/middleeast/cholera-yemen.html

Global Monitoring Laboratory. (2020). Trends in atmospheric carbon dioxide. Retrieved December 5, 2020, from https://www.esrl.noaa.gov/gmd/ccgg/trends/

Gofossilfree.org. (2020). *Gofossilfree.org*. Retrieved December 5, 2020, from https://gofossilfree.org/divestment/commitments/

Graham, L., Debucquoy, W., & Anguelovski, I. (2016). The influence of urban development dynamics on community resilience practice in New York City after Superstorm Sandy: Experiences from the Lower East Side and the Rockaways. *Global Environmental Change, 40*, 1112–1124. https://doi.org/10.1016/j.gloenvcha.2016.07.001

Green Bronx Machine. (2020). Green Bronx Machine. Retrieved December 5, 2020, from https://greenbronxmachine.org

Green Globe. (2020). Standard criteria and indicators. Retrieved December 5, 2020, from https://greenglobe.com/standard/

GRI. (2020). Welcome to GRI. Retrieved December 5, 2020, from https://www.globalreporting.org

Haass, R., Dittmer, P., Veigt, M., & Lütjen, M. (2015). Reducing food losses and carbon emission by using autonomous control—A simulation study of the intelligent container. *International Journal of Production Economics, 164*, 400–408.

Hardin, G. (1968). The tragedy of the commons. *Science, 162*, 1243–1248.
Heede, R., & Oreskes, N. (2016). Potential emissions of CO_2 and methane from proved reserves of fossil fuels: An alternative analysis. *Global Environmental Change, 36*, 12–20.
Henry, M., Beguin, M., Requier, F., Rollin, O., & Odoux, J. (2012). A common pesticide decreases foraging success and survival in honey bees. *Science, 226*, 348–350.
Higgins, R. (2007). *Thoreau and the language of trees*. University of California Press.
Hoegh-Guldberg, O., Mumby, P. J., Hooten, A. J., Steneck, R. S., Greenfield, P., Gomez, E., Harvell, C. D., Sale, P. F., Edwards, A. J., Caldeira, K., Knowlton, N., Eakin, C. M., Iglesias-Prieto, R., Muthiga, N., Bradbury, R. H., Dubi, A., & Hatziolos, M. E. (2007). Coral reefs under rapid climate change and ocean acidification. *Science, 318*, 1737–1742.
Hutner, H. (2018). Japanese women and antinuclear activism after the Fukushima accident. In R. Brinkmann & S. J. Garren (Eds.), *The Palgrave handbook of sustainability: Case studies and practical solutions* (pp. 283–298). Palgrave Macmillan.
IBM. (2020). GHG emissions inventory. Retrieved December 5, 2020, from https://www.ibm.com/ibm/environment/climate/ghg.shtml
ICLEI. (2020a). ICLEI local government for sustainability. Retrieved December 5, 2020, from https://www.iclei.org
ICLEI. (2020b). Greenhouse gas protocols. Retrieved December 5, 2020, from https://icleiusa.org/ghg-protocols/
IGF. (2020). IGF sustainability statement. Retrieved December 5, 2020, from https://www.igfgolf.org/sustainability/igf-statement-on-sustainability
Imperva. (2020). Bots. Retrieved December 5, 2020, from https://www.imperva.com/learn/application-security/what-are-bots/#:~:text=Social%20Media%20Bots,Twitter%20accounts%20are%20social%20bots
Interaction.org. (2020). The NGO Climate Compact. Retrieved December 5, 2020, from https://www.interaction.org/wp-content/uploads/2020/04/Climate-Compact.pdf
Iowa Corn. (2019). Corn production. Retrieved June 1, 2019, from https://www.iowacorn.org/corn-production/production/
ipbes. (2019). Media release: Nature's dangerous decline 'unprecedented'; species extinction rates 'accelerating'. Retrieved June 1, 2019, from https://www.ipbes.net/news/Media-Release-Global-Assessment

IPCC. (2014). *Climate change 2014: Synthesis report*. Retrieved June 1, 2019, from https://www.ipcc.ch/site/assets/uploads/2018/02/SYR_AR5_FINAL_full.pdf
IPCC. (2018). *Global warming of 1.5°C*. Retrieved June 1, 2019, from https://www.ipcc.ch/site/assets/uploads/sites/2/2018/07/SR15_SPM_version_stand_alone_LR.pdf
ISO. (2020). *ISO 26000—Social responsibility*. Retrieved December 5, 2020, from https://www.iso.org/iso-26000-social-responsibility.html
Jacobson, S., Pinto, J., Gutsche, R. E., & Wilson, A. (2019). Goodbye, Miami? Reporting climate change as a local story. In J. Pinto, R. E. Gutsche, & P. Prado (Eds.), *Climate change, media & culture: Critical issues in global environmental communication* (pp. 53–71). Emerald Publishing.
Karim, L. (2014). Disposable bodies: Garment factory catastrophe and feminist practices in Bangladesh. *Anthropology Now, 6*, 52–63.
Katner, A. L., Pieper, K., Lambrinidou, Y., Brown, K., Subra, W., & Edwards, M. (2018). America's path to drinking water infrastructure inequality and environmental injustice: The case of Flint, Michigan. In R. Brinkmann & S. J. Garren (Eds.), *The Palgrave handbook of sustainability: Case studies and practical solutions* (pp. 79–97). Palgrave Macmillan.
Klein, N. (2010). *The shock doctrine: The rise of disaster capitalism*. Picador.
Klein, N. (2014). *This changes everything: Capitalism vs. the climate*. Simon & Schuster.
Krakauer, J. (1997). *Into the Wild*. Villard.
Kretzschmar, P., Kramer-Schadt, S., Ambu, B., Bender, J., Bohm, T., Ernsing, M., Göritz, F., Hermes, R., Payne, J., Schaffer, N., Thayaparan, S. T., Zainal, Z. Z., Hildebrandt, T. B., & Hofer, H. (2016). The catastrophic decline of the Sumatran rhino (Dicerorhinus sumatrensis harrissoni) in Sabah: Historic exploitation, reduced female reproductive performance and population variability. *Global Ecology and Conservation, 6*, 257–275.
Krimmerer, R. W. (2013). *Braiding sweetgrass: Indigenous knowledge, and the teachings of plants*. Milkweed Editions.
Lam, M. K., Tan, K. T., Lee, K. T., & Mohamed, A. R. (2009). Malaysian palm oil: Surviving the food verses fuel dispute for a sustainable future. *Renewable and Sustainable Energy Reviews, 13*, 1456–1464.
Lappé, F. M. (1971). *Diet for a small planet*. Ballantine.
Larson, E. (2020, March 17). Exxon loses jurisdiction fight in Massachusetts climate suit. *Bloomberg Green*.

Learning Gate Community School. (2020). *Learning Gate Community School*. Retrieved December 5, 2020, from https://learninggate.org/

Leopold, A. (1949). *A Sand County almanac*. Ballantine.

Li, Y., Shi, Y., Zho, X., Cao, H., & Yu, Y. (2014). Coastal wetland loss and environmental change due to rapid urban expansion in Lianyungang, Jiangsu, China. *Regional Environmental Change, 14*, 1175–1188.

Louisville Open Data. (2020). *The urban heat island project*. Retrieved December 5, 2020, from https://data.louisvilleky.gov/story/urban-heat-island-project

LouisvilleKy.gov. 2020. *Sustainability*, Retrieved December 5, 2020 from https://louisvilleky.gov/government/sustainability

Lovelock, J. E., & Margulis, L. (1974). Atmospheric homeostasis by and for the biosphere: The gaia hypothesis. *Tellus, 26*, 2–10. https://doi.org/10.3402/tellusa.v26i1-2.9731

Lynch, M. J., Stetesky, P. B., & Long, M. A. (2016). A proposal for the political economy of green criminology: Capitalism and the case of the Alberta tar sands. *Canadian Journal of Criminology and Criminal Justice, 58*, 137–160.

Lynch, S. N., Lambert, L., & Bing, C. (2018, October 4). U.S. indicts Russians in hacking of nuclear company Westinghouse. *Reuters*. Retrieved June 1, 2019, from https://www.reuters.com/article/us-usa-russia-cyber/u-s-indicts-seven-russians-for-hacking-nuclear-company-westinghouse-idUSKCN1ME1U6?il=0

Maamoun, N. (2019). The Kyoto protocol: Empirical evidence of a hidden success. *Journal of Environmental Economics and Management, 95*, 227–256.

Mahapatra, S. K., & Ratha, K. C. (2016). Paris climate accord: Miles to go. *Journal of International Development, 29*, 147–154.

Markham, K. E., & Sangermano, F. (2018). Evaluating wildlife vulnerability to mercury pollution from artisanal and small-scale mining in Madre de Dios, Peru. *Tropical Conservation Science, 11*. https://doi.org/10.1177/1940082918794320

Markowitz, G., & Rosner, D. (2013). *Deceit and denial: The deadly politics of industrial pollution*. University of California Press.

Martin, P. S. (2005). *Twilight of the mammoths: Ice age extinctions and the rewilding of America*. University of California Press.

McIntosh, E. (2019, June 14). Michael E Mann took climate change deniers to court. They apologized. *Grist*. Retrieved June 14, 2019, from https://grist.org/article/michael-e-mann-took-climate-change-deniers-to-court-they-apologized/

Micklin, P. (2016). The future Aral Sea: Hope despair. *Environmental Earth Sciences, 75*. https://doi.org/10.1007/s12665-016-5614-5

Mielke, H. W., & Zahran, S. (2012). The urban rise and fall of air lead (Pb) and the latent surge and retreat of societal violence. *Environment International, 43*, 48–55.

Miller, D. (2012). *Last nightshift in Savar: The story of the spectrum sweater factory collapse.* McNidder & Grace.

Morelle, R. (2019, May 13). Mariana Trench deepest-ever sub dive finds plastic bag. *BBC News.* Retrieved June 1, 2019, from https://www.bbc.com/news/science-environment-48230157

Moussa, A. M. A. (2013). Predicting the deposition in the Aswan High Dam Reservoir using a 2-D model. *Ain Shams Engineering Journal, 4*, 143–153.

Murray, L. (2019). The need to rethink German nuclear power. *The Electricity Journal, 32*, 13–19.

Newsome, T., Aranguren, F., & Brinkmann, R. (1997). Lead contamination adjacent to roadways in Trujillo, Venezuela. *Professional Geographer, 49*, 331–341.

Odegard, I. Y. R., & van der Voet, E. (2014). The future of food—Scenarios and the effect on natural resource use in agriculture in 2015. *Ecological Economics, 97*, 51–59.

Our World in Data. (2020a). Energy. Retrieved December 5, 2020, from https://ourworldindata.org/energy

Our World in Data. (2020b). Renewable energy. Retrieved December 5, 2020, from https://ourworldindata.org/renewable-energy

Our World in Data. (2020c). Water use and stress. Retrieved December 5, 2020, from https://ourworldindata.org/water-use-stress

Oxford English Dictionary. (2020). Word of the year. Retrieved December 5, 2020, from https://languages.oup.com/word-of-the-year/2020/

Ozone Secretariat. (2010). The Montreal Protocol. Retrieved June 1, 2019, from https://web.archive.org/web/20130420100237/http://ozone.unep.org/new_site/en/Treaties/treaties_decisions-hb.php?sec_id=5

Packham, C. (2019, August 13). New Zealand PM Calls on Australia to answer Pacific island climate change demands. *Reuters.* https://www.reuters.com/article/us-pacific-forum-australia/new-zealand-pm-calls-on-australia-to-answer-pacific-island-climate-change-demands-idUSKCN1V327B

Pierre-Louis, K. (2017, March 28). What the frack is in fracking fluid? *Popular Science.* Retrieved June 1, 2019, from https://www.popsci.com/what-is-in-fracking-fluid/#page-2

Ramankutty, N., Foley, J. A., & Olejniczak, N. J. (2002). People on the land: Changes in population and croplands during the 20th century. *Ambio: A Journal of the Human Environment, 31*, 251–257.

Raptis, C. E., van Vliet, M. T. H., & Pfister, S. (2016). Global thermal pollution of rivers from thermoelectric power plants. *Environmental Research Letters, 11*. Retrieved June 1, 2019, from https://iopscience.iop.org/article/10.1088/1748-9326/11/10/104011

Reddy, C. M., Arey, J. S., Seewald, J. S., Sylva, S. P., Lemkau, K. L., Nelson, R. K., Carmichael, C. A., McIntyre, C., Fenwick, J., Ventura, G. T., Van Mooy, B. A. S., & Camilli, R. (2012). Composition and fate of gas and oil released to the water column during the Deepwater Horizon oil spill. *PNAS, 109*, 20229–20234.

Righter, R. W. (2005). *The battle over Hetch Hetchy: America's most controversial dam and the birth of modern environmentalism*. Oxford University Press.

Riva, M. A., Lafranconi, A., D'Orso, M. I., & Cesana, G. (2012). Lead poisoning: Historical aspects of a paradigmatic "Occupational Environmental Disease". *Safety and Health at Work, 3*, 12–16.

Roberts, H. H. (1997). Dynamic changes of the Holocene Mississippi River Delta plain: The delta cycle. *Journal of Coastal Research, 13*, 605–627.

Sasikumar, K. (2017). After nuclear midnight: The impact of a nuclear war on India and Pakistan. *Bulletin of the Atomic Scientists, 73*, 226–232.

Schumacher, E. F. (1973). *Small is beautiful: A study of economics as if people mattered*. Blond & Briggs.

Schwirplies, C. (2018). Citizen's acceptance of climate change adaptation and mitigation: A survey in China, Germany, and the U.S. *Ecological Economics, 145*, 308–322.

Sellers, C. C. (2012). *Crabgrass crucible: Suburban nature and the rise of environmentalism in twentieth-century America*. The University of North Carolina Press.

Sengupta, S., & Lee, C. W. (2020, February 13). A crisis right now: San Francisco and Manila face rising seas. *New York Times*.

Sessions, G. (Ed.). (1995). *Deep ecology for the twenty-first century: Readings on the philosophy and practice of the new environmentalism*. Shambhala Publications, Inc.

Sharaan, M., & Keiko, U. (2020). Projections of future beach loss along the Mediterranean coastline of Egypt due to sea-level rise. *Applied Ocean Research, 94*. https://doi.org/10.1016/j.apor.2019.101972

Shearer, C. (2011). *Kivalina: A climate change story*. Haymarket Books.

Shiva, V. (2016). *Staying alive: Women, ecology, and survival in India*. Zed Press.
Siegle, L. (2019, June 21). Fast fashion is on the rampage, with the UK at the head of the charge. *The Guardian*. Retrieved December 5, 2019, from https://www.theguardian.com/fashion/2019/jun/21/fast-fashion-is-on-the-rampage-with-uk-at-the-head-of-the-charge
Solar Impulse Foundation. (2020). Solar Impulse Foundation. Retrieved December 5, 2020, from https://solarimpulse.com
SRCDOM. (2020). Sustainable rural community development. Retrieved December 5, 2020, from https://www.betterplace.org/en/organisations/8219-sustainable-rural-community-development-surcod
Tampa Bay Water. (2020). Tampa Bay seawater desalination. Retrieved December 1, 2020, from https://www.tampabaywater.org/tampa-bay-seawater-desalination
Target. (2020). Climate. Retrieved December 5, 2020, from https://corporate.target.com/corporate-responsibility/planet/climate
TCdata360. (2020). Households w/ personal computers. Retrieved December 5, 2020, from https://tcdata360.worldbank.org/indicators/entrp.household.computer?country=BRA&indicator=3427&viz=choropleth&years=2016
Terrapass. (2019). Terrapass. Retrieved June 1, 2019, from https://www.terrapass.com/
The Minimalists. (2020). *The Minimalists*. Retrieved December 5, 2020, from https://www.theminimalists.com
The Nature Conservancy. (2019). Carbon calculator. Retrieved June 1, 2019, from https://www.nature.org/en-us/get-involved/how-to-help/consider-your-impact/carbon-calculator/?gclid=EAIaIQobChMIiY_MiqWs4QIVko7ICh1NnweYEAAYASAAEgKKb_D_BwE
Thunberg, G. (2018). *No one is too small to make a difference*. Penguin Books.
Tucker, W. C. (2012). Deceitful tongues: Is climate change denial a crime? *Ecology Law Quarterly, 39*, 831–894.
Unilever. (2020). Sustainable living. Retrieved December 5, 2020, from https://www.unilever.com/sustainable-living/
United Nations. (2020). Universal Declaration of Human Rights. Retrieved December 5, 2020, from https://www.un.org/en/universal-declaration-human-rights/
US Supreme Court. (2007). Massachusetts vs. EPA. Retrieved December 5, 2020, from https://supreme.justia.com/cases/federal/us/549/497/

Van den Akker, R., Gibbons, A., & Vermeulen, T. (Eds.). (2017). *Metamodernism: Historicity, affect, and depth after postmodernism*. Rowman & Littlefield Publishers.

van Wyk, J. (2018). Political leadership and sustainability in Africa: Margaret Kenyatta. In R. Brinkmann & S. J. Garren (Eds.), *The Palgrave handbook of sustainability: Case studies and practical solutions* (pp. 347–358). Palgrave Macmillan.

Vercoe, R., & Brinkmann, R. (2012). A tale of two sustainabilities: Sustainability in the global north and south to uncover meaning for educators. *The Journal of Sustainability Education*. http://www.susted.com/wordpress/content/a-tale-of-two-sustainabilities-comparing-sustainability-in-the-global-north-and-south-to-uncover-meaning-for-educators_2012_03/

Warren, K. J. (1997). *Ecofeminism: Women, culture, nature*. Indiana University Press.

Watson, I., Shelley, J., Pokharel, S., & Daniele, U. (2019, April 27). China's recycling ban has sent America's plastic to Malaysia. Now they don't want it—So what next? *CNN*. Retrieved June 1, 2019, https://www.cnn.com/2019/04/26/asia/malaysia-plastic-recycle-intl/index.html

Weller, Z. D., Hamburg, S. P., & von Fischer, J. C. (2020). A national estimate of methane leakage from pipeline mains in natural gas local distribution systems. *Environmental Science and Technology, 54*, 8958–8967.

Wilkins, D. (2020). Catholic clerical responses to climate change and Pope Francis's Laudato Si'. *Environment and Planning E: Nature and Space*. https://doi.org/10.1177/2514848620974029

Wootson Jr., C. R. (2018, July 7). YouTube daredevils filmed dangerous stunts for clicks—Then died going over a waterfall. *The Washington Post*. Retrieved December 5, 2020, from https://www.washingtonpost.com/news/the-intersect/wp/2018/07/07/youtube-daredevils-filmed-dangerous-stunts-for-clicks-then-died-going-over-a-waterfall/

World Bank. (2019a). Fertilizer consumption. Retrieved June 1, 2019, from https://data.worldbank.org/indicator/ag.con.fert.zs

World Bank. (2019b). Concept project information document (PID)—Egypt: Greater Cairo air pollution management and climate change project—P172548 (English). Retrieved December 5, 2020, from https://documents.worldbank.org/en/publication/documents-reports/documentdetail/622831576851407904/concept-project-information-document-pid-egypt-greater-cairo-air-pollution-management-and-climate-change-project-p172548

World Bank. (2020a). *What a waste 2.0: A global snapshot of waste management to 2050.* Retrieved December 5, 2020, from https://www.worldbank.org/en/news/infographic/2018/09/20/what-a-waste-20-a-global-snapshot-of-solid-waste-management-to-2050

World Bank. (2020b). Greater Cairo air pollution management. Retrieved December 5, 2020, from https://projects.worldbank.org/en/projects-operations/project-detail/P172548?lang=en

Index

A

AASHE, see Association for the Advancement of Sustainability in Higher Education
Abbey, Edward, 76, 77
Adventure tourism, 254–256
Afghanistan, 255
Africa, 8, 42, 46, 100, 166, 183, 202, 206, 212, 256, 270
Agenda 21, 264
Agricultural waste, 178, 181, 182
Airport, 46, 55, 248
Air quality, 26, 217, 251, 272
Alaska, 22, 47, 255, 256
Alberta, 43, 165, 166
Algae, 41, 42
Amazon, 46, 58, 100, 136, 202, 203, 219
Amsterdam, 15, 244
Anderson, Ray, 54
Animal rights, 79–81
Anthropocene, 39–40, 45, 61
Anthropocentrism, 68–69
Apartheid, 48, 126
Appalachian Mountains, 40
Aquatic ecosystem, 193
Aquifer, 44, 159, 164, 166, 167, 270
Aral Sea, 44, 73, 159, 161, 164
Asia, 45, 142, 166, 183, 196, 199, 201
Association for the Advancement of Sustainability in Higher Education (AASHE), 125, 126
Asthma, 217
Aswan High Dam, 152
Athens, 44
Australia, 13, 14, 18, 19, 23, 32, 33, 45, 61, 125, 153, 166, 179, 201, 206, 236
Authenticity, 8

B

Bahamas, 233, 236, 244, 248
Balkans, 212
Bangladesh, 61
Beach, 9, 191, 193, 216, 244
Beirut, 178
Belize, 249
Benchmark/benchmarking, 26–28, 62, 135, 136, 224, 230, 231, 251, 254
Black, 47, 210, 217
BlackRock, 28, 29, 127, 136, 235, 238, 239
Bloomberg, Michael, 26
Blue baby syndrome, 42
Bog, 46, 200, 204
Bolsonaro, Jair, 18, 33, 203
Bone Valley Formation, 165
Borat, 6
Boreal forest, 193
Boy Scouts of America, 129
Brazil, 18, 31–33, 54, 58, 151, 153, 202, 203, 206, 252
British Columbia, 256
British Petroleum (BP), 134
Brownfield, 26
Buddhism, 227
Bullard, Robert, 56, 57, 75
Business, 3, 10, 24, 27–29, 33, 52–55, 59, 62, 81–83, 85, 109, 110, 114, 121, 122, 132–133, 141, 146, 163, 179, 222, 228–231, 233, 236, 237, 241, 243, 244, 252, 253, 269, 273
Buy Nothing Day, 187, 190

C

Cairo, 104, 106–108, 198
California, 44, 109, 163, 166
Cambodia, 212
Canada, 25, 52, 58, 151
Capitalism, 51–53, 79, 227, 228, 234
Car, 4, 8, 14–16, 55, 90, 91, 96, 97, 107, 117, 132, 177, 188, 195, 196, 247, 248, 258, 266
Carbon, 4, 16, 29–33, 52, 54, 61, 88, 92–94, 97–100, 104, 106, 119, 126, 130, 134, 136, 155, 165, 247, 258
Carbonate rock, 31
Carbon credit, 87, 93, 100–101, 117, 126, 155, 247, 258
Carbon cycle, 32, 39, 160
Carbon dioxide, 16, 17, 31, 32, 88, 93, 98, 100, 144, 145
Carbon dioxide equivalent, 16, 17, 26, 88–94, 100, 101
Carbon footprint, 10, 30, 85, 87–99, 121–133, 246, 248, 259
Carbon sink, 87, 93, 100–101
Cardosa, Fernando Henrique, 203
Carson, Rachel, 74, 273
Caves, 32
Cenozoic Era, 165
Chernobyl, 148
Chicago, 5, 38, 245
Chicken, 98, 234
Childish Gambino, 6
Chile, 52, 152
China, 24, 25, 59, 91, 142–144, 151–153, 165, 179, 183, 227, 248
Chlorofluorocarbon, 17, 25
Chuckchi Sea, 22
Citizens Campaign for the Environment, 270
Clean Air Act, 73
Clean Water Act, 73, 168

Climate, 11, 13, 15–21, 23, 26, 33, 40, 44, 78, 88, 89, 101–104, 107, 117, 127, 129, 132, 134–136, 141, 144–147, 193, 198, 201, 238, 244, 254, 263, 270, 273, 275
Climate action plan, 26, 27, 29–32
Climate change, 1, 3, 5, 7, 8, 10, 11, 13–34, 37, 46, 51, 52, 61, 68, 73, 76, 78, 82, 85, 87, 94, 101–110, 121, 123, 124, 128–132, 134, 136, 141, 143, 151, 153, 156, 161, 201, 204, 228, 230, 237–239, 256, 263, 264, 270, 273
The Climate Pledge, 136
Coal, 14, 16, 26, 29, 33, 47, 50, 71, 89, 104, 117, 134, 141–144, 147, 150
Colorado River, 161
Community, 3–6, 8, 21–23, 26–28, 32, 37, 38, 40, 44, 50, 53, 54, 56–62, 69, 72, 81, 83, 85, 98, 99, 103–118, 121, 124, 128, 129, 136, 141, 146, 151, 154, 163, 164, 166, 167, 170, 172, 174, 181, 183, 189, 191, 201–203, 205, 207, 210, 211, 216, 217, 219, 220, 222–224, 230, 234, 236, 240, 241, 245, 246, 252, 253, 256, 257, 262–264, 268–274
Consumerism, 11, 99, 186–188, 190, 229, 261–266
Consumption, 7, 8, 14, 45, 48, 54, 55, 89, 98, 135, 142, 143, 147, 156, 162, 163, 187, 205, 228, 234, 253, 265
Contaminants, 43, 167–170, 182

Convention on International Trade in Endangered Species of Wild Fauna and Flora (CITES), 73
Cooling, 16, 28, 87, 89, 90, 93, 94, 133, 162, 198
Coral reef, 10, 193, 195, 200, 201, 206, 250, 269, 270
COVID, 6, 21, 38, 52, 98, 133, 180, 246
Cozumel, 248

D

Dakota Access Pipeline, 58, 272
De Blasio, Bill, 26
Dead zones, 42, 169, 205
Deep ecology, 76–78
DeKalb, 167, 199, 245
Delta, 40, 42, 73, 107, 169, 192
Denial, 18–19
Denton, 30
Desert, 164, 193, 195, 218, 252, 255
Developed nations, 6, 180, 185, 222
Dhaka, 61
Diamond, 48
Diet, 45, 74, 124
Dirty energy, 134, 141–147, 150, 155, 156
Door County, 244
Dorsey, Tom, 246
Dust Bowl, 72
Duterte, Rodrigo, 23

E

Earth, 3, 16, 17, 39, 40, 46, 48, 61, 68, 78, 79, 101, 102, 147, 151, 153, 165, 167, 179, 193, 249, 250

Earthquake, 71, 148
Earth Summit, 24, 264
Ecofeminism, 78–79
Ecological footprint, 4, 6, 11, 45, 246, 250, 267
Ecology, 78, 164, 173, 198, 202, 252, 274
Economic development, 38, 52–54, 58, 122, 129, 228, 231–234, 241, 244, 256, 272
Economy, 4, 8, 10, 14, 24, 33, 38, 45, 51–55, 61, 83, 123, 132, 183, 184, 223, 227–241, 246
Ecosocialism, 79
Ecosystem, 3, 10, 16, 20, 21, 30, 32, 37, 40, 45–47, 50, 61, 68, 70, 71, 73–77, 79, 100, 139, 151, 152, 161, 164, 165, 179, 191–205, 250, 252, 255–257, 269
Ecotourism, 135, 245, 256–258
Edison, Thomas, 142
Education, 6, 11, 53, 76, 106, 122–126, 196, 207, 212, 221, 233, 241, 251, 262, 266, 269–271
Egypt, 104, 106, 107, 152, 153, 218, 244
EJSCREEN, 220, 222
Electric car, 4, 54, 97, 247, 258, 271
Electricity, 6, 23, 50, 71, 89, 94, 96, 142, 143, 145, 147, 150–154, 183
Electronic waste, 51, 58, 60, 135, 178–180, 182
Endangered species, 41, 45, 68, 73, 80, 195, 196, 269
Endangered Species Act, 73

Energy, 6, 7, 10, 16, 18, 22, 24–26, 28–30, 33, 38, 41, 45–47, 50–54, 61, 89, 90, 92, 94–99, 101, 106, 111, 112, 114, 115, 117–119, 123, 125–127, 129, 134–136, 139, 141–155, 182–184, 198, 219, 228–230, 232–234, 236, 237, 241, 247, 248, 251, 252, 266, 269, 273
Environment, 1, 4, 8–11, 16, 24, 28, 32, 34, 38, 40–43, 49, 51, 54, 60–62, 67–71, 73, 74, 76–79, 82, 83, 88, 92, 107, 123, 125, 127, 154, 169, 171, 173, 178, 182, 183, 193, 194, 197, 198, 200, 207, 230, 233, 236, 237, 243, 244, 247, 248, 252, 253, 255, 257, 258, 268, 270, 274, 275
Environmental ethics, 1, 9, 67, 69, 72, 76, 82, 83
Environmental justice, 11, 26, 38, 56–58, 60–62, 211, 219–224, 264, 269, 272
Environmental Protection Agency (EPA), 16, 57, 109, 110, 113, 115, 162, 168, 169, 219, 220, 222, 251, 273
Environmental racism, 1, 3, 4, 11, 38, 56, 62, 75, 207, 211, 217–219, 224
Environmental sustainability, 10, 38, 40–51, 55, 124, 139
EPA, *see* Environmental Protection Agency
Equity, 4, 6, 8, 10, 34, 38, 54, 56–62, 83, 211, 230, 252
Estuary, 193, 195

Ethics, 72–74, 81–84, 185, 261, 266–268
Europe, 98, 142, 143, 148, 179, 183, 212, 231
European, 25, 41
European Union, 52, 94, 143
Eutrophication, 41, 73, 76, 170, 205
Everglades National Park, 70, 200
Extinction, 40, 45, 46, 79, 194, 203, 275
Extinction Rebellion, 78
Exxon, 18, 22
Exxon Valdez, 47

Facebook, 4, 262–264, 270
Factory, 16, 29, 61, 186
Fairbanks, 255
Farmers, 46, 55, 58, 98, 164, 165, 195, 203
Fashion, 49, 80, 166
Fast fashion, 49, 184, 185, 187, 265
Fen, 46
Fertilizer, 41, 165, 168, 169, 171, 174, 204, 252, 267
FGBC, *see* Florida Green Building Coalition
Fink, Larry, 28, 238
Flint, 57, 62, 75, 168
Flint River, 57
Flood, 21, 129, 151, 263
Florida, 27, 42, 70, 103, 123, 165, 166, 170, 191, 192, 195–197, 200, 244, 246, 250, 251, 274
Florida Green Building Coalition (FGBC), 27
Florida panther, 70, 195, 196

Food, 4, 20, 27, 29, 38, 43, 45, 51, 53–55, 58, 75, 81, 87, 89–91, 93, 98, 99, 101, 106, 124–126, 135, 162, 165, 181, 189, 190, 196, 202, 205, 217, 218, 230, 233, 234, 241, 244, 252, 265, 266, 272
Food deserts, 218
Food Not Bombs, 99
Forest, 31, 58, 77, 109, 184, 193, 195, 198, 201–203
Fortune 500, 27, 55, 235, 236
Fossil fuel, 8, 16, 18, 24–26, 29, 31, 33, 47, 48, 88, 126, 127, 136, 142, 147–149, 172, 183, 230, 237, 238, 241, 250, 264
Fracking, 47, 145
Fukushima, 148, 168

Garbage, 10, 51, 57, 58, 107, 112, 114, 115, 118, 177–189, 245
Germany, 24, 26, 33, 142, 152, 153
Glaciers, 161, 167, 257
Globalization, 52, 57, 60–62, 109, 186, 222, 237
Global Reporting Initiative (GRI), 83
Goldman Sachs, 239
Golf, 71, 117, 169, 171, 192, 252–254, 258
Gore, Al, 68
Government, 3, 5, 6, 10, 14, 24, 27, 52, 54, 58, 72, 73, 76, 77, 82, 103, 104, 108–110, 113, 114, 116–118, 128, 149, 163, 164, 168, 178, 196, 200, 205, 212, 217, 219, 222, 231–233, 238, 239, 264, 272

Grassland, 191, 193, 256
Great Britain, 142, 244
Great Plains, 72
Green, 4, 11, 26, 27, 33, 38, 41, 53–55, 67, 117, 127, 132, 135, 136, 149–151, 155, 156, 162, 173, 227–241, 261, 264, 266, 271, 273
Green energy, 7, 8, 10, 14, 26, 30, 33, 115, 117, 123, 141–155, 230, 232–234, 236, 237, 241, 248, 269, 273
Green Globe Certification, 135
Greenhouse effect, 15, 16
Greenhouse gas, 10, 14, 16, 17, 19–21, 24–33, 55, 83, 85, 87–89, 91–93, 96–118, 121–123, 126, 128, 132–136, 141, 145–147, 161, 247, 273
Greenhouse gas calculator, 89
Greenhouse gas equivalent, 93
Greenhouse gas footprint, 10, 88–93, 100, 101
Green New Deal, 26, 38
Greenpeace, 67
Groundwater, 23, 42, 44, 45, 47, 50, 58, 73, 161, 167, 168, 170, 174, 182, 200, 252, 270
Gulf of Mexico, 40, 42, 47, 169

H

Hardin, Garrett, 72
Hawaii, 17
Hazardous waste, 50, 56, 62, 178, 179, 182, 220
Heating, 16, 87, 89, 90, 93–96, 106, 133, 143, 145, 148, 198, 265

Heavy metals, 10, 50, 153, 179, 183, 195
Helium, 48
Hetch Hetchy, 71, 72
Highway Beautification Act, 177
Hill, Julia Butterfly, 77
Hoarders, 186
Honeybee, 205
Housing, 23, 26, 29, 71, 89, 130, 190, 220
Human rights, 11, 83, 163, 164, 207, 211–216, 223, 224
Hurricane, 21, 85, 191, 192
Hydraulic fracking, 47
Hydrocarbon, 43, 47
Hydroelectric power, 47, 150, 151

I

IBM, 133
ICLEI, 27, 28, 30, 109, 110, 113, 115, 264
Identity, 6
Illinois, 77, 167, 199, 245
India, 24, 25, 94, 142, 153, 179
Indiana, 245
Indonesia, 31, 202
Industrial waste, 178, 180–182
Infrastructure, 7, 21, 23, 24, 30, 37, 46, 96, 97, 117, 118, 146, 155, 163, 168, 178, 244, 247, 248
Innuit, 22
Insurance, 29
Intergovernmental Panel on Climate Change (IPCC), 8, 16, 17, 19, 21, 23, 109
International Organization for Standardization (ISO), 83, 222, 223

IPCC, *see* Intergovernmental Panel on Climate Change
Iran, 255
ISO 26000, 222–224
Italy, 103, 198, 232

J

Japan, 148, 153, 179, 183

K

Kazakhstan, 44, 73, 159
Kentucky, 104, 105, 119, 183
Kenyatta, Margaret, 270
Kivalina, 22, 23
Klein, Naomi, 51, 52, 228
Kyoto Protocol, 24–26

L

Lake Okeechobee, 191, 192, 196
Landfill, 16, 30, 50, 60, 99, 101, 144, 182–185, 188
Las Vegas, 37, 44, 244, 252
Latinx, 210, 217
Latvia, 244
Leadership, 5, 11, 26, 33, 55, 82, 104, 110, 116, 117, 122, 136, 266, 268, 269, 271
Lead pollution, 209, 210
Lebanon, 178
Leissner, Tim, 239
Leisure, 243–259
Leopold, Aldo, 74
Lesser developed nations, 5
LGBTQ, 216
Lifestyle, 4, 11, 12, 40, 45, 55, 61, 88, 91, 97–99, 187–190, 230, 239, 262, 266, 267, 270, 271, 275
Lions Club, 128, 129
Liquid waste, 173, 178, 180
Litter, 7, 60, 170, 177, 180, 182, 248, 250, 268
London, 41, 142, 198, 235
Los Angeles, 44, 93, 198, 247
Louisiana, 40
Louisville, 5, 104–108, 119
Luxembourg, 45
Luxor, 243, 244

M

Madonna, 6
Manila, 22, 23
Mann, Michael, 17, 263
Mao, 6
Margarita Island, 244
Marine ecosystem, 152, 193
Marsh, 46, 193, 200
Marshall Plan, 231
Massachusetts, 18, 110
Mass transit, 96, 97, 117, 118, 218
Matrix, The (film), 6
Mauna Loa, 17
Mauna Loa observatory, 17
McCandless, Christopher, 255
Meat, 41, 45, 46, 79, 80, 92, 123, 134, 181, 182
Medical waste, 178, 180, 182
Mercury, 42
Mesozoic Era, 16
Metamodernism, 6–8
Methane, 16, 17, 30, 88, 93, 98, 99, 101, 145, 182
Methemoglobinemia, 42
Mexico, 152, 153, 248, 249

Michigan, 57, 62, 75, 168, 232
Microplastic, 171
Microsoft Teams, 133
Middle East, 145, 183
Milan, 198, 232
Millennium Development Goals, 33
Milwaukee, 171, 199
Minimalist, 55, 187–189, 272
The Minimalists, 188, 272
Minerals, 48, 165
Mining, 40–43, 48, 71, 142, 147, 149, 153, 162, 165, 166, 181, 187, 202
Minnesota, 21
Mississippi River, 40, 42, 73, 169
Missouri River, 58
Modernism, 6
Monroe, Marilyn, 6
Montreal Protocol, 25
Morrison, Scott, 18
Muir, John, 71
Mumbai, 23
Museum of the North, 255

N

Nanoparticles, 168, 172
Nassau, 248
National Organization of Women, 129
Natural gas, 7, 16, 33, 47, 50, 88, 89, 94, 112, 117, 143–147, 204
Nature, 3, 11, 39, 52, 68–72, 74, 76–78, 82, 89, 104, 114, 146, 149, 171, 205, 206, 223, 230, 249, 251, 253, 255, 256, 268, 275
Neoliberal, 52, 60, 62, 228, 237

Nevada, 37
New York, 5, 18, 26, 38, 91, 93, 98, 112, 163, 183, 198, 233, 235, 247
Nile, 107, 152, 249
Nitrate, 42, 169, 178
Nitrogen, 40, 41, 74, 168, 169
Non-profit, 3, 10, 24, 82, 85, 109, 110, 121–133, 205, 233, 257, 269, 273
North America, 53, 125, 142, 179
North American Free Trade Act, 52
Norway, 26, 152
Nuclear energy, 147–149
Nuclear waste, 149
Nuclear weapons, 149
Nutrient, 10, 40–43, 123, 167, 169–171, 192, 195, 201, 202
Nutrient pollution, 41, 161, 201

O

Obama, Barack, 54
Obesity, 217
Ocean, 19, 20, 41, 43, 50, 60, 100, 159, 161, 167, 168, 192, 193, 195, 201, 249
Ocean dumping, 50
Offsets, 29–32, 92, 136
Ogallala Aquifer, 44, 164
Oil, 7, 22, 43, 46, 47, 58, 89, 117, 134, 143–145, 147, 168, 173, 181, 202, 204, 230, 249, 250
Old Testament, 68
OneNYC, 26
Organic, 6, 16, 53, 54, 88, 99, 167–169, 178, 181, 182, 200, 204, 229, 233, 234, 237

Organic food, 4, 53, 54, 230, 234, 241
Ozone, 25, 74, 220

P
Pacific, 13, 19, 23
Pacific Island, 13, 33
Palm oil, 46, 202
Paris, 125, 245
Paris Climate Accord, 14, 24–26, 28, 32, 76
Park, 26, 27, 56, 69–72, 191, 198–200, 205, 234, 252, 255, 271
People for the Ethical Treatment of Animals (PETA), 80, 81
Permafrost, 203, 204
Personal care products, 27, 75, 80, 172
Peru, 165
Pesticide, 10, 167, 174, 204, 205, 252, 267
Petroleum, 16, 32, 33, 43, 47, 58, 134, 141–145, 150, 166, 210
Pharmaceuticals, 10, 74, 168, 171, 172
Philippines, 22, 23, 59, 96
Phoenix, 93
Phosphate, 165, 166
Phosphorus, 40, 41, 169
Photovoltaic cells, 153
Pinchot, Gifford, 72
Pipeline, 16, 58, 143, 145
PlaNYC, 26
Plastic, 7, 8, 10, 41, 43, 51, 58–60, 75, 171, 180, 181, 184, 185, 188, 189, 195

Pollutant, 10, 43, 74, 110, 168–172, 174, 177, 183, 195, 209, 217
Pollution, 3, 10, 19, 20, 24, 26, 29, 38, 41–44, 47, 48, 56, 60, 61, 68, 73–76, 106–108, 123, 139, 142, 143, 149, 161, 169, 170, 172–174, 177, 178, 181, 199, 201, 209, 210, 217, 227, 228, 248–250, 252, 264, 270
Pope's Climate Covenant, 82
Ports, 46, 179, 191, 248–250
Post modernism, 6
Power plant, 16, 26, 50, 111, 142, 144, 145, 147–149, 153, 162, 168, 183, 219
Prairie, 77, 193
Preservation, 69–72, 80, 205, 233, 234, 256, 269
Public health, 6, 7, 10, 21, 108, 149, 170, 183, 209, 210
Public land, 76, 269
Pyrenean ibex, 46

Q
Qatar, 45

R
Racism, 1, 38, 56, 62, 75, 207, 211, 217–219, 224, 275
Rainforest, 10, 46, 58, 68, 100, 201–203, 206, 219
Rainforest Alliance, 206, 241
Rare earth minerals, 48
Razak, Najib, 239
Recycle, 8, 50, 59, 179, 181, 185

Recycling, 7, 51, 53, 54, 58–60, 73, 149, 179–182, 184, 185, 188, 189, 201, 251, 264
Redwood Summer, 109
Reef, 31, 193, 195, 200, 201, 206, 249, 250, 257, 269, 270
Renewable energy, 6, 10, 33, 53, 54, 134, 136, 149, 150, 155, 156, 234
Resource Conservation and Recovery Act, 177
Resources, 3, 37, 38, 40, 41, 44, 45, 48, 49, 52, 68, 72, 78, 102, 108, 110, 125, 130, 136, 142, 149, 159–173, 210, 216, 227, 228, 232–234, 253, 257, 272
Riga, 244
River, 40, 42, 43, 57, 58, 61, 71, 123, 151, 152, 159, 161, 167–169, 174, 198
Roads, 46, 96, 117, 177, 220, 247
Rock, 40, 47, 145, 160, 200
Rock cycle, 40, 160
Rome, 244
Russia, 52, 152

S

Salt marsh, 193
Sana'a, 44, 166
San Francisco, 44, 71, 198
Sanitary landfill, 50, 182
Sanya, 248
Saudi Arabia, 153, 163, 252
Savanna, 193
Schools, 10, 76, 85, 109, 110, 121–133, 221, 268, 274
Schumacher, E. F., 54, 55, 227, 228

Science, 1, 6, 14–19, 68, 102, 264, 274
Sea ice, 22
Sequestration, 29–32
Sewage, 42, 74, 112, 115, 118, 144, 170, 171, 180, 181, 184, 249, 250
Shanghai, 23, 250
Shiva, Vandana, 78
Sierra Club, 71
Snow, John, 170
Social media, 261–267, 269
Social networks, 21
Soil, 44, 98, 100, 101, 170, 191, 193, 195, 201, 202, 209, 210
Soil erosion, 72, 73, 100
Solar, 8, 16, 30, 33, 94, 96, 115, 117, 123, 132, 143, 150, 153–155, 198, 229, 247, 266, 271
Solar Impulse, 247
Solid waste, 26, 177
South Africa, 48, 126, 237
South America, 42, 183
Spain, 153, 166
Species, 3, 40, 41, 45, 46, 68, 73, 79, 80, 195–198, 203, 269
Sri Lanka, 52
Stakeholders, 29–31, 53, 62, 82, 105, 117, 121, 126, 130, 133, 136, 219, 220, 222, 223, 231, 233, 235, 252, 269, 270, 273
Standing Rock Sioux, 58
STARS, 126
Subsidence, 23, 45
Suburbanization, 46
Sulfur, 40
Sustainabillies, 55

SUV, 4
Swamp, 16, 46, 195, 199, 200

T

Tampa, 93, 123, 170
Tampa Bay, 166
Tampa Bay Water, 166
Target, 93, 108, 115, 134, 253, 273
Tar sands, 42, 43, 165, 166
Teams, 4, 19, 31, 50, 97, 122, 230, 255, 269
Tel Aviv, 44
Temperature, 15–17, 20, 21, 94–96, 106, 161, 193, 198, 201, 204
Terrapass, 101
Tesla, 8, 101, 247
Tetra Pak, 30, 185
Texas, 30, 233
Thirty-Day Sustainability Challenge, 4
This is America, 6
Three E's, 4, 8, 38, 40, 51, 54, 83, 229–231
350.org, 127
Thunberg, Greta, 33, 78, 98, 129, 250, 270, 271
Tiger King, 79–80
TikTok, 264
Tourism, 11, 134, 135, 207, 233, 243–249, 254–258
Transportation, 4, 26, 28, 29, 58, 87, 89–91, 93, 96–98, 101, 107, 117, 118, 130, 141, 143, 146, 186, 190, 198, 245, 246, 248, 269
Travel, 11, 90, 91, 98, 101, 112, 117, 126, 133, 135, 207, 243–259, 266

Tropical rainforest, 10, 46, 201–203
Tropical storm, 13, 21, 23
Trump, Donald, 6, 18, 33, 54, 204, 263
Tundra, 31, 100, 193
Turkey, 153
Turkmenistan, 162
Tuvalu, 13, 14
Twitter, 262–264, 267

U

Unilever, 27, 28, 230
United Arab Emirates, 153
United Kingdom, 54
United Nations (UN), 8, 9, 27, 46, 73, 83, 98, 109, 212, 235, 264
United States (USA), 5, 18, 21–26, 30, 32, 33, 38, 42, 44–47, 52–54, 56–62, 69, 73, 81, 82, 93, 94, 96, 104, 109, 110, 125, 142, 143, 151–153, 161–164, 166, 168, 177–189, 191, 195, 200, 207, 210, 217, 219–222, 230, 233, 244, 264
U.S. Green Building Council (USGBC), 273
U.S. Green Chamber of Commerce, 230
Universal Declaration of Human Rights, 212–216
Universities, 122, 123, 125–127, 234, 237, 238, 255, 273, 274
University of Alaska, 255
University of California, 238
Urban ecology, 198
Urban heat island, 106, 198

USGBC, *see* U.S. Green
    Building Council
US Forest Service, 72
Uzbekistan, 44, 73, 159

V

Vacation, 11, 88, 91, 97, 101, 161,
    243, 247, 252, 255,
    258–259, 266
Vatican Bank, 238
Vegan, 99, 267
Vegetarian, 91, 99
Vehicle, 4, 26, 29, 98, 115, 117,
    132, 146, 220, 232, 247, 264
Venezuela, 52, 227, 244
Venice, 103
Verizon, 136

W

Walmart, 52, 53, 55, 62, 229, 230
Warhol, 6
Waste, 7, 8, 10, 26, 27, 30, 41, 43,
    48–51, 56, 58–60, 62, 74, 87,
    90, 91, 98, 99, 101, 115, 135,
    139, 149, 165, 170, 173,
    177–185, 187–189, 195, 205,
    220, 223, 228, 233, 251,
    252, 266
Waste management, 50, 59, 73, 99,
    114, 153, 182–184, 189, 252
Waste to energy, 30, 50, 118, 183,
    184, 219
Wastewater, 30, 75, 112, 114, 115,
    166, 170, 171, 220, 221
Water, 3, 14, 16, 27, 37, 38, 40–44,
    47, 48, 57, 58, 71, 73–75,
    94–96, 123, 126, 145, 148,
    149, 151, 152, 159–173, 179,
    182, 188, 191, 192, 196, 198,
    200, 201, 205, 216,
    250–252, 271–273
Water cycle, 40, 160–161, 174
Water quality, 6, 47, 62, 159,
    166–171, 250, 270
Water supply, 23, 26, 57, 71, 75,
    164, 167, 168
Waterway, 26, 41, 73, 161, 181,
    244, 251, 252, 275
Weather, 17, 18, 20, 151, 161, 247
Wetland, 46, 73, 76, 100, 159, 161,
    170, 191, 192, 194, 197, 199,
    200, 204, 244
Wilson, Woodrow, 72
Wind, 8, 17, 33, 96, 115, 123, 143,
    150, 152–155, 250
Wisconsin, 14, 244
Women, 27, 78, 129, 217, 218, 250
World Bank, 41, 50, 108

Y

Yemen, 37, 44, 96, 166
Yosemite National Park, 71
Young Evangelicals for Climate
    Action, 82
YouTube, 256, 267
Yucatan, 248

Z

Zero-waste, 99, 190
Zoning, 117, 155
Zoo, 79, 80
Zoom, 133, 235

Printed by Printforce, the Netherlands